Edwin Lincoln Moseley

Sandusky flora

A catalogue of the flowering plants and ferns growing without cultivation, in Erie

County, Ohio

Edwin Lincoln Moseley

Sandusky flora
A catalogue of the flowering plants and ferns growing without cultivation, in Erie County, Ohio

ISBN/EAN: 9783744739832

Printed in Europe, USA, Canada, Australia, Japan

Cover: Foto ©berggeist007 / pixelio.de

More available books at **www.hansebooks.com**

Ohio State Academy of Science.

SPECIAL PAPERS NO. 1.

SANDUSKY FLORA.

A CATALOGUE

OF THE

FLOWERING PLANTS and FERNS

GROWING WITHOUT CULTIVATION, IN ERIE COUNTY, OHIO,
AND THE PENINSULA AND ISLANDS
OF OTTAWA COUNTY,

By E. L. MOSELEY, A. M.

PUBLISHED BY THE ACADEMY OF SCIENCE,

MAY, 1899.

PRESS OF CLAPPER PRINTING CO.

WOOSTER, OHIO.

TO THE MEMORY

OF THE

MEMBER OF THE ACADEMY

WHOSE DEATH IS ANNOUNCED,
AS THE PROOF OF THE LAST PAGES OF THIS "SPECIAL PAPER"
ARE BEING RETURNED TO THE PRINTER,

MANNING F. FORCE.

GENERAL, JURIST, SCIENTIST, AND ABOVE ALL, A MAN WHOM
NO DESIRE FOR WEALTH OR FAME COULD DIVERT
FROM THE FAITHFUL SERVICE OF HIS
FELLOW MEN, THIS WORK IS

DEDICATED.

SANDUSKY FLORA.

The flora of the Sandusky district is a rich one. We believe there is no other local collection of Ohio plants that approaches within three hundred species of the number collected in the past seven years, in Erie county and the eastern part of Ottawa county, and now preserved at the Sandusky High School. Of the many local lists published in other states, we have seen none that give so many native species as have been found near Sandusky, although several of them cover much larger areas and represent the labors of many botanists working for long periods of time. Some of these lists, moreover, include territory that is regarded especially rich in plants.

The "Flora of Buffalo and its Vicinity," by David F. Day, presents the names of all the plants which have been detected within fifty miles of Buffalo, a territory many times as large as Erie county, Ohio, and including on the one side the whole of the Niagara river with its profusion of flowers and ferns, and, on the other mountains with an altitude of 2300 feet above the sea. "The Cayuga Flora" by William R. Dudley, published as a Bulletin of the Cornell University, covers an area 65 miles in extreme length and is based on numerous collections, the first of which was made in 1827. The "Plants of Monroe county, New York, and Adjacent Territory," published by the Rochester

Academy of Science, in addition to Monroe county, which is about three times as large as Erie county, Ohio, includes portions of five other counties and gives twenty species reported by early botanists, but no longer found. All of these districts border on Lake Ontario and one of them on Lake Erie also.

The whole of England contains but about 1200 native phenogams; surpassing the little district about Sandusky by less than a hundred species.

Although several hundred native plants not found in Erie county grow in one place or another in Ohio, yet so well is the flora of the state represented here, that it is probably not too much to say that excepting the counties bordering on the Ohio river and those that contain sphagnous swamps or bogs, there are few counties in the state where a botanist, unfamiliar with the territory would be likely to find in a single day's search more than half a dozen native species that do not grow somewhere in Erie county. The surpassing richness of the Sandusky flora is not due to the fact that it includes islands within its territory, for scarcely any of its species are confined to the islands, nor is it in very large measure due to the fact that it includes species that are confined to the lake shore but rather to peculiarities of climate and geological features, both of which depend to some extent on the proximity of the lake.

CLIMATIC INFLUENCE OF LAKE ERIE ON VEGETATION.

The Catalogue of Canadian Plants in six volumes includes the whole territory lying north of the Great Lakes and extending from the Atlantic to the Pacific. The Sandusky district contains 165 native species and varieties not given in the Canadian catalogue besides a

number of others which in Canada are confined to Pt.
Pelee or Pt. Pelee Island, spots only a few miles distant
from the islands of Ottawa and Erie counties, Ohio.
The Sandusky district contains 67 native plants not
known to grow anywhere in Michigan and many
others which in Michigan are confined to the south-
western part where the climate is tempered by Lake
Michigan. But what seems quite as remarkable is the
fact that the Sandusky district contains 305 native
plants not known to grow within fifty miles of Buffalo,
while the Buffalo district has about 244 native species
and varieties not given in the Sandusky list. But even
this great difference between two regions bordering on
Lake Erie is largely due to climate, for the summer at
Buffalo is not only cooler but lasts less than three-
fourths as long as at Sandusky. Since the prevailing
winds have traversed Lake Erie for nearly its whole
length before reaching Buffalo the mean temperature in
summer there is about 3° lower than at Sandusky. In
the spring the difference is even greater than in summer,
being about 5¼° lower in April and May. This is due
to the fact that when the ice breaks up it is blown to
the east end of the lake and remains so crowded there
as to prevent navigation three weeks or more after
Sandusky Bay has been clear. The average date of the
last killing frost in spring in Sandusky, is April 14; at
Buffalo, May 20, that is 36 days later. Moreover, Sandusky is protected by its position from cold north-west
winds in autumn, while Buffalo is not, so that the first
killing frost at Sandusky does not come on an average
until October 23, but at Buffalo October 5, that is
18 days earlier.

Like an east and west mountain range, Lake Erie
protects the plants on its south side from cold north
winds while they get the full force of winds from the
south, but with the vegetation on the north side it is
the reverse. Moreover, the heat given out by the

water in winter as it freezes, modifies the climate of the
adjacent land. It would seem that an equal amount
of heat should be absorbed by the ice in melting, and
thus the winter prolonged into spring, but for the
region about the western end of the lake this is not
true, because a great part of the ice is blown away
toward the east end of the lake, whose period of cold
is prolonged thereby. And so it comes that the climate
on the south side of Lake Erie is not only milder than
that on the north side but much milder than that at
the east end, and, if we reckon the length of summer
from the average date of the last killing frost in
spring to the average date of the first killing frost in
autumn, we find the summer at Sandusky to last 192
days and at Buffalo only 138 days.

The counties of Ohio lying to the east of Erie
county and bordering on the lake have a climate some-
what less mild than that of the Sandusky region for
their land rises more abruptly from the water, and the
prevailing winds pass over more of the lake before
reaching them. In Erie county the land within a few
miles of the lake is mostly much less than a hundred
feet in elevation. The temperature at Sandusky in
spring and summer averages about one and a half
degrees higher than at Cleveland, and one degree higher
than in the eastern part of Erie county, four miles back
from the lake shore where Mr. W. H. Todd has recorded
observations for the government for many years.

It is interesting to observe that the protection from
frost afforded by Lake Erie scarcely extends beyond the
counties that border upon it and, as a result we have
many plants in these that have not been reported from
any other county north of the middle of the state, and
quite a number that have been found nowhere else in
Ohio except in the southern part, within forty miles of
the Ohio River. Even so far south as Columbus, the

last killing frost in spring occurs on an average six days later than at Sandusky and the first killing frost in autumn five days earlier.

CLIMATE OF THREE CITIES

ON LAKE ERIE AND ONE A HUNDRED MILES SOUTH OF IT FROM TIME OF ESTABLISHMENT OF WEATHER BUREAU IN EACH PLACE TO END OF 1897.

		Sandusky.	Cleveland.	Buffalo.	Columbus.
WEATHER BUREAU ESTABLISHED.		1878.	1869.	1870.	1878.
MEAN TEMPERATURE.	January............	26.2	26	25	28.4
	February..........	29.4	26	25.3	32.1
	March.............	34.7	33	30.5	38.1
	April...............	47.7	46	42.5	51.2
	May............	59.5	58	54.2	62.0
	June.................	68.8	68	65.4	71.3
	July...	73.6	72	70.2	74.9
	August.............	71.6	70	69.1	72.3
	September..	65.6	64	62.5	66.1
	October.....	53.7	52	51.2	53.7
	November........	41.2	40	38.4	41.2
	December..........	32.8	31	30.5	33.3
	Annual...	50.4	48.8	47.1	52.1
Lowest Minimum.		−16	−17	−14	−20.3
Highest Maximum		100	99	95	103
Av. date last killing frost in spring		April 14.	May 1.	May 20.	April 20.
Av. date 1st killing frost in fall.		Oct. 23.	Oct. 11.	Oct. 5.	Oct. 18.
Av. rainfall in inches.....		34.69	34.82	39.66	38.74
Av. relative humidity......		72.07	72.0	74.5	71.4

GEOLOGY.

The physical feature of Erie county which causes most difference between its flora and that of the counties to the east is the existence of prairies in its southern and western part. These prairies are of two sorts, each having its characteristic plants, while many species not known to grow farther east in the state are found on both of them.

Extending over the greater part of the township of Oxford, and over portions of the townships of Milan, Huron, Perkins, Margaretta and Groton is a nearly level prairie which probably at one time formed the bottom of the glacial lake that preceded Lake Erie and later of a bay or bays partly shut off from the lake by sand bars which still exist. Underlying most of this prairie is the Ohio shale, which in many places is close to the surface. The ground requires tiling to produce good crops. The other prairie lies north and west of the village of Castalia, extending to the western boundary of the county. The soil of this is different from the other, being a calcareous deposit from the water of the Castalia springs. Within the memory of men still living a great deal of this prairie was under water much of the time. A considerable portion of the region extending south of Castalia for a distance of over fifty miles has no surface streams, but the water descends through the joints of the limestone and flows through subterranean passages which it has made in the soluble rock of the Waterlime formation. This water charged with lime carbonate issues from the ground in numerous bold springs in the vicinity of Castalia, which owes its name to this circumstance. These springs are the largest and most beautiful in Ohio. The slope from Castalia to Sandusky Bay is very gradual and before any artificial drainage was established, the region was a marsh filled more or less with the calcareous water whose deposits have formed over thousands of acres to a depth of many feet. In some places these deposits are indurated forming a tufa, in others, soft making a shell marl containing the remains of millions of Limnea and Planorbis of the same species as live in the bay now. The tufa is composed mostly of petrified Chara and other plants. The shape and venation of leaves is well preserved, one of the most common kinds being that of Hypericum kalmianum which grows over much of the surface. On this prairie as well as on the Oxford prairie grow many plants not found east of the Huron river either in Erie county or the counties beyond.

Sandusky and Margaretta as well as Marblehead Peninsula and Kelley's Island are underlaid by Corniferous limestone which comes near the surface over much of this region. In many places, especially on Marblehead, the covering of soil is only a few inches or a fraction of an inch deep and consists of partially decomposed vegetation and lime carbonate derived from the underlying rock. Quite a number of species are characteristic of this calcareous soil. Catawba Island, as it is called, and the islands of the Put-in-Bay group have a similar character but the rock is older, belonging to the Waterlime formation. Over the greater part of Sandusky and in many places on the islands, the limestone is covered with clay of variable thickness, but in many parts the soil is too thin for trees to attain a large size, for even if they could obtain nourishment enough, they are likely to be uprooted by a strong wind. The glacier that passed over this region left traces that still show in hundreds of places, including some grooves on Kelley's Island and Marblehead which so far as we know are unsurpassed elsewhere in the world. It is interesting to observe that the grooves on the different Islands, on the Peninsula and in Sandusky and Margaretta have the same direction, running about twelve degrees south of west, or parallel with the axis of Lake Erie, excepting a few which have quite a different direction and indicate a movement of the ice at a different time. Where the superimposed drift has protected the rock from weathering, it not only retains the deep grooves but shows everywhere a highly polished surface marked with fine parallel lines.

In many places in Sandusky this polished limestone requires no quarrying to serve admirably for basement floors. So level is the rock and the overlying drift that for miles around the city, the eye can scarcely detect any elevatioas or any depressions with the exception of slight ones made by small streams.

Many of the rare plants of Erie County grow in sand, especially in the sand deposits east of the village of Milan and along the sand ridges that stretch east and west in Margaretta township and along the border of the prairie in the southern part of Perkins township. These were formerly lake beaches and just below the sand ridge that extends south-west from Castalia is a ledge of limestone which shows very plainly the action of the waves, though it is now four miles from the water. When the lake had settled to a lower level, it must have beat against the foot of this ledge, undermining the rock and causing it to break away in large masses, as it is doing now at the west end of Rattlesnake Island and elsewhere. These detached masses often settled but a few feet, leaving deep but narrow chasms between them and the parent cliff, and these chasms are but partially filled even to the present day with dirt washed in from above. In places, trees grow out of them and the walls are bedecked with ferns. The rich woods covering the side of this hill, which I have called Margaretta Ridge, the sandy fields at the top and the prairie below afford a variety of plants found nowhere else in the county and a large number of species unknown in the counties farther east.

The Huron River divides Erie county into an eastern and western part. Few of the plants which grow in Erie county and not in Lorain or Cuyahoga counties are found east of this river. West of it are no natural surface streams that continue to flow all summer and except near the river no ravines. The valley of the Huron and its tributaries therefore afford some species not found nearer Sandusky, but as it is cut through shale, it is not so rich as the valleys farther east. At Berlin Heights, the Old Woman Creek has cut a picturesque ravine through the Waverly sandstone and into the Ohio shale. Here grow several interesting plants not found farther west. But still deeper ravines have been formed in Florence township by the Vermillion River and its tributaries, the walls mainly of shale, but in the southern part of the township also of sandstone. Here have been found many species of sedges and other plants that do not seem to grow along the Huron or west of it, though most of them grow in the counties to the east where there are still deeper ravines in the sandstone. The walls of these ravines like the walls of a cellar are warmed slowly in summer, so that on the north side of steep, wooded slopes, are some cooler spots than any near Sandusky and hence many plants which are more common farther to the north and east.

The lake shores and marshes furnish quite a number of species not found in the interior of the state. Cedar Point consists of low sand ridges thrown up by the lake and separating it from Sandusky bay and its marshes for a distance of seven miles. Throughout most of its length the plants are comprised in few species but toward the end it is wider and probably older, having a richer soil and more varied flora. Although more accessible from Sandusky than any other good collecting ground and appearing not to have a great number of species, yet so many rare forms grow there in one place or another that it is not improbable that some plant not on our list at all may yet be found there. Seven years ago, before the work of making a herbarium had been commenced, the writer thought he had found on Cedar Point about all the species that grew there, but each year he has added

something from that region, which he had never found
before either there or elsewhere. In the number of rare
species, Cedar Point is surpassed by Marblehead,
though the latter has a larger area. Altogether the
Sandusky district has furnished more than a hundred
species and varieties that were not known to be growing wild any where in the state, previous to their
discovery here.

FLORA OF THE ISLANDS AND ITS ORIGIN.

With the exception of some of the little ones, the
islands of Ottawa county, and Kelley's—the only
island belonging to Erie county,—have been visited
many times and at different seasons. Of the plants
growing on six of the islands in the lake, separate lists
have been kept and an attempt made to make them
complete. These lists are not published except as a
part of the general list of plants comprised in the
Sandusky flora, but a fair idea of the results may be
obtained from this by bearing in mind that all the
plants marked common or abundant have been found
on one and, in nearly all cases, on more than one of the
islands, except a few which are noted otherwise. The
names of plants not common on the mainland but
occuring on Kelley's Island and two or more of the
Put-in-Bay group are followed by the word—Islands.
If found on Kelley's island and only one other, or not
on Kelley's the names of the islands on which the
plant has been found are given. The number recorded
for each island is as follows:

> Kelley's Island.............................461.
> Put-in-Bay................................439.
> Middle Bass..... ....306.
> North Bass................................282.
> Rattlesnake...............................192.
> Green Island..............................115.

It will be seen that the numbers correspond pretty well with the size of the islands, the largest island having the greatest number of species, the next in size the next greatest number, etc. The different islands are very similar in character. consisting of limestone covered more or less with clay and without any permanent streams. The difference in physical features and the difference in flora between the islands are much less than between parts of the mainland of Erie county separated by shorter distances. The entire number of different species is 612. Fourteen of these are Naiadaceae growing in the water of bays or along the shore, most of them at Put-in-Bay and North Bass. The islands are poor in ferns, the whole number of species being only eight, of which Kelley's has six, Put-in-Bay three, all scarce, Green Island two, Middle Bass and Rattlesnake one each, and North Bass none. We have found on them no orchids and no Ericaceae. Kelley's island, owing to its extensive commerce and cultivation, together with the protection from frost afforded by the water, has many naturalized species, especially along the south shore, two or three of which have not been noticed elsewhere. Excepting these and three rare sedges, and one rare golden rod, the islands appear to have no plants that have not been found also on the mainland of Erie county or on Marblehead, —not so many species as are afforded by each township of Erie county, excepting Groton. However, in view of the fact that the islands have no permanent streams, no ravines, no alluvial soil and little or no sand except the barren sand in some places along the shores, their flora is probably as varied as that of equal areas on the mainland where these defects exist. Their combined area is only about ten square miles.

It has been supposed that the lake, which after the melting of the southern portion of the glacier overspread a larger area than Lake Erie does now, sub-

sided until what are now the islands appeared above
its surface. This view is doubtless correct, but there is
now much evidence to show that it continued to
subside until the islands formed part of the mainland
and afterward rose and isolated them again, and is
still rising and likely to submerge them again. The old
beaches which may be traced for long distances
running nearly parallel to the present shores of the
Great Lakes, must have been level at the time they
were formed, but they are not now level, and there has
therefore been a tilting of that part of the earths crust
which includes the basins of the Great Lakes, as there
has been of many other parts. These beaches all have
gentle slopes, toward the south and south-west, indi-
cating that in this part of North America, there has
been an uplifting of the land toward the north
north-east or a depression toward the south
south-west or both. The effect of this tilting
of the basins of these lakes has been to
raise the water on the south and west as
compared with that on the opposite sides, just as the
tipping of a saucer partly filled with water would do.
The fluctuation of the water due to variable winds and
rainfall make such comparisons difficult, but Mr. G. K.
Gilbert found by comparing the heights above the
normal level of Lake Erie in 1895, of a certain point in
Cleveland, and a certain point at the head of the Wel-
land canal with the heights of the same two points as
carefully determined in 1858, that the point near the
north-east end of the lake rose 0.239 foot as compared
with the point in Cleveland. This is a small amount
and in view of the difficulty of determining the normal
level and measuring the exact height of any point on
the land above it even by measurements many times
repeated, it might well be attributed to some inaccur-
acy in the measurements if it were an isolated case. but
it is not. Similar comparison of points on Lake

Ontario and on Lake Huron and Michigan also, indicate tilting, and tilting in the same direction as at Lake Erie and not only that but the amount corresponds with the distance apart of the two points compared. Furthermore the direction of the tilting indicated by these measurements is the same as that indicated by the dip of the old lake beaches. We are therefore forced to the conclusion that the basins of the great lakes have been considerably tilted and that this tilting has been going on in the present century. As the outlet of Lake Erie is at that end of the basin which has been raised more than any other part, the result has been to deepen the water throughout, but especially at the opposite end where the islands are situated. The spreading of the waters over the land should be here more noticeable for another reason also, viz.; because the shores are so low. We should therefore expect to find here in the form of submerged forests and other things that could not have formed under water, evidence of the spreading of the waters of the lake over the land, and so we do.

OLD TREES KILLED BY RISE OF THE WATER.

By the high water that prevailed in 1858 to 1860 large trees were killed in many places where the waves could not reach them. Mr. George Hine, who owns land bordering the marsh east of Sandusky, had hickory trees two feet in diameter killed in this way. On Kelley's Island large sycamore trees standing on the border of the south marsh, on Put-in-Bay elm and sycamore, on Middle Bass big trees growing by the marsh near Rehberg's, and at Toussaint and elsewhere along the shore between Port Clinton and Toledo old walnut trees, were killed at this time by high water keeping the ground too wet around their roots. Persons who came to Erie county in the forties remember seeing about the marshes connected with the bay many

dead trees which they believed to have been killed by
high water, and old residents of Put-in-Bay and
Kelley's Island have told me the same thing about
trees there. It is probable that these trees were killed
in 1838 when the water was nearly as high as in 1858,
though it did not remain high so long. Hundreds of
walnut stumps are still standing along the border of
the marshes east of Sandusky where even now, although
the water is lower than usual, it is too wet for walnut
trees to grow. One that stood recently on ground only
six inches above the present lake level measured 5 feet 4
inches in diameter. We may infer from this that during
the life of this tree, probably over three hundred years,
the water was not so high as in the present century.

SUBMERGED FORESTS.

Stumps and logs with roots attached have been
found under water and show that when the trees grew
the water must have been considerably lower than it
has been during the present century. In the lake at
Deisler's bathing beach, Put-in-Bay, was a sycamore
stump that was dangerous to persons swimming, as it
did not show above the water, and had to be blasted
out. Other stumps in the water not far from where this
one stood may still be seen. Near the Black Channel in
Sandusky bay are cedar stumps standing upright with
roots in place and completely submerged, except at
such low stages of the water as rarely occur, when a
little of the tops project. About a mile west of Venice
many buried cedar stumps have been found below the
level of the lake.

Besides stumps a large amount of submerged timber
that fell without being cut has been found where it fell,
and much of it is to be seen now. The greatest
quantity is in the Huron marsh connected with Sandusky Bay. In parts where the water and mud are not
very deep the logs may be easily seen in such numbers

and variety as to show that a forest was once there, but in the deeper water they are also abundant and are often struck by the pole of a hunter pushing his boat through the marsh. When in a very dry season, the ditch was dug through the marsh in order to float boats from the club house out to open water, logs of sassafras with the roots on, and a cedar with branches were found at the bottom, *i. e.* 3 or 4 feet below the present lake level. Even in the deeper parts a few logs are still to be seen partly above the water, having been supported by roots, or roots and branches until the marsh had grown up under them. A cedar out about 60 rods from land where the muck is five feet deep, has roots extending down into it at least three feet. It is 17 inches in diameter, and has about 60 rings. A pine log two feet in diameter and with 91 rings lies where the muck is over six feet deep. It has roots running down some distance and 30 years ago was not yet prostrate but the other end stuck up as much as seven feet above the water, and formed a landmark for fishermen. This is out about 80 rods from the present shore of the marsh. A walnut tree that forks into two huge and crooked branches whose ends are buried in the muck must have grown near where it lies, but this also, though a mile or more from the pine log, is some 80 rods out from shore, and the water at this place is now seven feet deep. It is still 23 inches in diameter and probably required nearly two centuries to grow. Observations on these trees were made March 5th and 6th, 1898, when the readings of the water gauge at Cleveland show the lake to have been 3½ feet lower than the high water mark. During the life of these trees the lake must have been at least eight feet lower than it has been during much of the time for the last forty years.

A great quantity of submerged timber still retaining roots and branches was removed from the water in front of the club house on Put-in-Bay by Mr. Vroman. There were soft maple, oak and sycamore, some of the logs four or five feet in diameter.

SUBMERGED STALAGMITES.

In several of the caves at Put-in-Bay nearly half a mile from shore, is deep water which rises and falls with fluctuations in the level of Lake Erie. The floors of these caves are covered with stalagmites, and the roofs were formerly studded with stalactites. In three caves I have seen stalactites hanging down in the water and in two stalagmites rising in the water. In one cave about thirty stalagmites may be seen on a submerged floor of a few square rods extent. They are, most of them, nearly cylindrical in shape, and represent merely the cores of larger stalagmites which once probably formed a crust over the whole floor, the remainder having been dissolved away. Those in the deeper water appear to have been dissolved more than those in the shallower parts. Many were standing in water from 2½ to 3½ feet deep, March 12 and 13, 1898. As stalactites and stalagmites would not form under water, the water from which the calcium carbonate was precipitated to form them must have flowed to a lower level than where the lowest stalagmites now exist. We may therefore infer that during the period of their formation, which certainly lasted many years, and probably some centuries, the lake was at least five feet lower than the mean level of the past forty years.

If these caves were formed in preglacial times, the argument still holds good, for if the lake had been as high or higher than now ever since the melting of the glacier and stalagmites had existed in the caves then, they would have been dissolved long ago. The stalagmites visible now are evidently not preglacial. Where the water does not cover the floor of the caves they are forming at the present time.

RIVER CHANNELS BELOW THE LAKE LEVEL.

In the Huron marsh off the mouth of Plum Brook, a setting pole may be pushed down 12 feet. This may be done along a line extending from the mouth of the creek out into the marsh, but a few rods on either side the pole goes down only two or three feet. When the stream cut this channel Lake Erie must have been at least 12 feet lower. Not only has the lake spread its waters over all the lowland through which this creek formerly flowed, and other creeks, whose submerged channels could doubtless be found by searching, but it has extended far up into the valleys of all the streams. This effect must result from the rise of the lake, for the streams had cut their valleys below the general level of the country, though not below the level to which the water had to flow while the cutting was going on. The Portage, the Sandusky, the Huron, and the other so-called rivers as well as all the smaller streams that enter this part of the lake, have the lower portions of their valleys filled by the water of the lake. Into the valley of the Old Woman Creek the lake has extended two miles farther than the present shore line, into the valley of the Huron five miles measured in a straight line from the present shore, into the Sandusky 22 miles beyond the Cedar Point light house, and more than 25 miles measured in a straight line from Rye Beach, for it is probable that the Black Channel at the east end of what is now Sandusky Bay is a part of the old river channel, also that the "Harbor" between Marblehead and Catawba is part of the old valley of the Portage, the lake having spread over the land to the west of Catawba and made an opening for the river at Port Clinton. This is not as yet quite certain, but there is no uncertainty about the valley of the Huron; it is still uninterrupted from the village of Huron on the lake shore to the place five miles inland where the flowing stream meets the water of the lake. The valley was

cut by the river when its waters continued to descend
to Huron and beyond, but this must have been when
the lake was not less than 32 feet lower than now, for
the bottom of the channel is 32 feet below the present
lake level at a point more than four miles from the lake,
and the depth of the water above the mud is between
17 and 32 feet all the way from this place to the lake.

Even Mud Creek, a small tributary of the Huron,
has all the lower part of its channel deep below the
present lake level. The entire drainage area of this
creek is only about four square miles, yet its waters
reach the present level of the lake nearly a mile
measured along the valley of the stream above its
junction with the Huron, and at a bridge about three-
fourths mile up the valley the water and mud are 12 or
14 feet deep.

EVIDENCE OF THE WATER'S DEEPENING IN THE
PRESENT CENTURY.

Records of the lake level kept at different places
show that at four times in the first half of the century
the water was lower than at any time in the last half.
In 1810 and in 1819 it was lower than any time since
1820, in 1841 and 1846 lower than at any time since
the latter date. In the absence of any record of exact
measurement of lake levels west of Cleveland we have,
nevertheless, evidence that the water about Sandusky
and the islands was lower in the early part of the
century. Mr. Shook, now living at Port Clinton,
remembers that in 1828 Mr. Ramsdell made hay of the
wild grass that grew on what is now the harbor west
of Lakeside, and that there was very little water then
where it has since been four feet deep. Similar state-
ments are made by other persons regarding this and
other places in this region.

When Harrison's army passed near Huron in 1813

a corduroy road about 60 rods long was built across
Mud Creek bayou, which, it is said, had been submerged
for many years, when, in 1867, the water being tempo-
rarily very low, Mr. Carpenter removed many of the
logs.

A survey made in 1887 of the Huron marsh at the
east end of Sandusky bay shows that a tract of land
one-half mile square, surveyed in 1809 has since become
marsh with the water and mud 12 to 18 inches deep,
and for two miles west of it, as far as it was surveyed,
the shore line has moved south about five rods. These
changes are certainly *not* due to erosion. Elsewhere
about Sandusky bay and along the shore of the lake
land has disappeared, partly from erosion and partly
because of the rising water covering it and giving the
waves new points of attack. The western part of the
city of Sandusky has suffered much from the encroach-
ment of the bay and along nearly the whole shore west
to Martin's Point and beyond land has disappeared.
So it is also along the lake. The surveys show that
for seven miles west from the Vermillion River the lake
has encroached upon the land between 20 and 34 rods
since 1809. From the Huron River to Dr. Esch's place,
about one and one-half miles west, the shore line has
moved south a distance varying from 18 to 28 rods,
west of this not so much. Since 1809 more than 500
acres have been lost to Erie county along the lake and
in the eastern part of the bay, and many acres more
between Sandusky and the western limit of the county.
On the north side of the bay, too, the water has
extended, open water covering ground where cat-tails
once grew. John Stone of Put-in-Bay, and Warren
Smith of Sandusky, remember when rushes grew over
much of Sandusky Bay where now is open water.
Until the middle of the century an island known as
Peninsula Point extended across nearly the whole
breadth of what is now the mouth of the bay. For the

length of a mile its height was 20 to 25 feet or more, and along the west side was clay covered with six inches of black soil bearing shell bark hickory trees and white oaks two and one-half feet in diameter. The last of this large island disappeared in 1860.

Gull Reef, north of Kelley's Island, has for many years been the greater part of the time under water. As late as 1850 it was an island on which stood a fish shanty and a tree that probably took a hundred years to grow.

DERIVATION OF THE ISLAND FLORA.

The facts stated in the preceding paragraphs suggest the possibility of many of the plants now on the islands having spread over them when a land connection existed between them and the mainland. Mr. Gilbert and others have concluded from a study of the old lake beaches that when the melting of the ice to the north opened an outlet for the glacial lake at Niagara the waters went down till it occupied only one-sixth the area that Lake Erie does now, and extended no farther west than Erie, Pa. We have seen that the submerged forests and stalagmites in the region about Sandusky and the islands prove a lower condition of the water when these were formed than has existed in the present century, and that the submerged river channels in this region indicate that the depression of the land as compared with the water has amounted to not less than 32 feet. A lowering of the water 22 feet would make it possible to walk from Kelley's island to Catawba, and 30 feet from Put-in-Bay to Catawba, excepting for a narrow channel, like a river which is deeper than the rest. We would be entitled, therefore, to conclude, even without a knowledge of observations made in other regions, that the islands were connected with the mainland in postglacial times. With this conclusion it is much easier to harmonize the facts ascertained regarding the plants now growing on the islands than to see how all of them could have been transported across several miles of water.

The seeds of many plants are provided with such means of transportation as would render their safe passage over a few miles of water an easy matter. Some produce fruit that is swallowed whole by birds and the pulp digested but not the seeds. The latter may thus be transported over land or water and propagate the species miles away from the parent plant. A mountain ash found growing on Rattlesnake Island in a thicket where birds roost was doubtless carried there in this manner. Some seeds like those of thistle have down so light that the wind may carry them long distances. Some are capable of floating for a time and then germinating. Some seeds are so small that they are likely to be carried in the mud that sticks to tne feet of rails or other birds that frequent marshy places. In several instances a single specimen of orchid has been found growing on some springy bank or damp place in the woods of Erie county and not another of the same kind within many miles. In two instances the single specimens are the only ones we have ever found in the county. These probably came from seeds that stuck to the feet of woodcocks or other birds that transported them from some distant bog. *Ammania coccinea* and some other mud-inhabiting species were probably transported in this way to the shore of Sandusky bay from much farther south for they are not known to grow elsewhere within more than a hundred miles.

When the ice forms a bridge between the islands and the mainland it would secm that weeds or their seeds might be blown across it or be carried across in the hair of animals. Seeds might also have been transported in former times by the Indians in their boats. In the present century the flora of the islands has been materially increased through the agency of man. Several cultivated plants have run wild and become well established there, including several species which are seldom found flourishing in the wild state so far north. The islands seem to have their full share of weeds and most of these have probably been introduced with impure seed. Others have probably been transported in baled hay and in packing material, and some, like the hore-hound, by sticking to people's clothes.

So numerous are the ways in which seeds may be transported that it would seem quite possible for the islands in the course of a few thousand years to have acquired all the plants that grow on them without any closer connection with the mainland than now exists. When, however, we consider more carefully these means of transportation in relation to all the species on the islands, we find it difficult to understand how some of them could have reached the islands in any of these ways.

A tornado passing first over the land and then the islands might carry seeds of any sort, but it would require more than one tornado to distribute seeds to all the islands and if any of the islands owed part of their plants to this agency we should expect to find on them some species well distributed which do not grow on the other islands at all, but this is not the case, with the exception of some species recently introduced by man. Other winds would not be likely to carry so far any but the lightest of seeds. Violent winds coming from the south where the mainland is nearest are generally accompanied by rain.

Any plant whose seeds are safely transported in the alimentary canal of birds might reach the islands in this way. Of the species that grow in muddy or marshy places and produce small seeds likely to be transported in mud on the feet of woodcocks, etc., not many occur on the islands and some of the islands have no places which such birds frequent.

Men who have often crossed the ice in winter say it would be impossible for seeds to be blown along on the ice all the way to the islands. Not only is the ice apt to be rough in many places, but it is crossed by numerous drifts of snow and is always intersected by long cracks in which seeds would lodge. Cakes of floating ice might transport seeds some distance, but would usually be prevented from landing them on distant shores by other ice getting in the way, and the freezing of the seeds to the floating ice would prevent them from blowing off. However, some littoral species may have reached the islands in this way. In those instances in which animals have succeeded in swimming so far, any seeds that were clinging to their hair at the start would probably be washed off on the way. Yet many species that rely upon mammals for transportation from place to place are there and give evidence of having been there longer than civilized man. These plants mature their seeds from four to six months before the ice would permit an animal to cross to the islands, and some of them have lost all their seeds by that time.

The following list gives the names of some of the plants on the islands whose seeds are adapted to transportation in the hair of animals: *Desmodium canescens, Desmodium paniculatum, Agrimonia eupatoria, Geum album, Geum virginianum, Circaea lutetiana, Osmorrhiza brevistylis, Osmorrhiza longistylis, Sanicula marylandica, Sanicula marylandica var. canadensis, Galium aparine, Galium boreale, Galium circaezans, Galium triflorum, Coreopsis trichosperma var. tenuiloba, Echinospermum virginicum.*

Colonel James Smith in the narrative of his captivity with the Indians, 1755–59, says: "These islands are but seldom visited; because early in the spring and

late in the fall it is dangerous sailing in their bark
canoes; and in the summer they are so infested with
various kinds of serpents, (but chiefly rattlesnakes,)
that it is dangerous landing." It is not probable then
that the Indians planted anything there, or that any
great number of seeds were introduced by them
accidentally.

The difficulty of seeds floating to the islands is two-
fold. The prolonged soaking in the absence of definite
currents to carry them in that direction is sufficient to
destroy the vitality of many kinds. The shores of the
islands do not afford conditions suited to the growth
of many of the species found in the interior. On Green
and Rattlesnake islands there is not a single spot
where it seems possible for a plant to start from seeds
washed ashore, except such as grow on bare rocks. Six
kinds of oak and three of hickory grow on the islands.
If all these kinds came from nuts that drifted ashore,
one would expect to find somewhere on the shore of
some island a tree so situated as to suggest the possibility of its having originated in this way, but not a
single one has been found. These are long lived trees,
and if within the period represented by the growth of a
large oak or hickory, there has not been a single in-
stance of a nut drifting ashore and finding a suitable
place to grow it may well be doubted, if in several
thousand years there would be opportunities for all the
different kinds to reach so many different islands. The
fact that acorns left in the water soon lose their power
to germinate increases the difficulty, yet it is not easy
to see how, except by floating, acorns or pig-nuts
would be likely to reach the islands as long as they
were separated from the mainland as far as they are
now.

The weeds that have followed civilized man from
the Old World, or have spread since the cultivation of the land from other parts of this, grow on

the islands as well as the mainland. That they have reached the islands mainly through man's agency is shown by the fact that those islands which have the most extensive commerce have the greatest variety of weeds. Green Island, being still wild, may be left out of consideration, but the greater part of Rattlesnake is cultivated, and there many kinds of weeds grow with a luxuriance that tries the patience of the owner. Yet there are fourteen kinds of weeds that grow on all four of the other islands, which are not to be found on Rattlesnake, without counting a number that need a damper soil than there prevails. Not only are most of these fourteen common on all the islands that enjoy much commerce, but among them are included a number of the most abundant weeds in this part of North America. The list is as follows: *Lepidium virginicum, Abutilon avicennæ, Melilotus alba, Medicago lupulina, Bidens frondosa, Sonchus asper, Xanthium canadense var. echinatum, Marrubium vulgare, Amarantus albus, Amarantus blitoides, Acalypha virginica, Juncus tenuis, Bromus secalinus, Panicum sanguinale.* Why are these species, elsewhere so abundant, not represented on Rattlesnake Island? For many years the island has been cultivated and the conditions suitable to the growth of these fourteen kinds of weeds, most of which have abounded for many years all around Lake Erie, but the island has been the abode of only a single family and its commerce, therefore very limited, and the seeds have not found any way to reach the island, or, if they floated to it, no way to get up onto soil where they could grow.

If a large portion of the plants on the islands have reached them in ways which may be called accidental and not by means that may be seen operating in the present century, then we ought to find deficiencies in the flora of certain islands due to the failure of certain species to reach them. Some plants that are well distributed on certain islands should be altogether wanting on others where the conditions for their growth are just as suitable. Moreover we should expect to find that some species not adapted to passing over the water had failed to reach any of the islands. But what we do find is the reverse. Every native species that is well distributed in similar soil on the mainland grows also on the islands and in no case, we believe, is a native species common over one island and lacking on others where similar conditions exist.

The leading facts bearing on the origin of the island flora may be summarized as follows: Within the present century the waters of Lake Erie and of the bays and marshes connected with it have encroached upon the land in the vicinity of Sandusky, covering many hundreds of acres of what was, at the time of the first surveys, solid ground. Trees several centuries old have been killed by high water in the present century. Submerged forests have been found in different parts of the region, submerged stalactites and stalagmites in the caves of Put-in-Bay, and submerged river valleys both east and west of Sandusky. When the trees grew and the stalagmites and valleys were formed, the land must have been above the level of the lake. The valleys are now deeper below the surface of the lake than is the lake bottom between the islands and the mainland. At the time they were formed. therefore, the lake did not separate the islands from the mainland. The flora of the islands is different from what we should expect to find, if all the species growing there had reached them by being transported across the water. It is probable then that many species have been on the islands since a time when these formed part of the mainland.

We may picture to ourselves woods such as grow at Lakeside now stretching north to Put-in-Bay and Kelley's island, interspersed here and there with prairies, perhaps, like those on the Peninsula now. We may well believe the picture to represent what was once a reality. How long ago this was we cannot tell. Some observations make it seem probable that it was not a great many centuries ago, perhaps less than twenty. Sometime we may find better means of judging.

SOUTHERN AND WESTERN PLANTS WHICH GROW NEAR LAKE ERIE.

Owing to the long summer enjoyed by places situated on the south shore of Lake Erie, many plants grow here which are not found farther north. As the country farther east lacks prairies such as occupy a considerable part of Erie county, quite a number of species appear to reach their eastern limit here. Since a number of the species are both southern and western, no separation of southern and western species is attempted in the following list. Many of the southern species grow east of the southern part of Lake Michigan. and some of them in southern Minnesota, where the summer isotherms reach a higher latitude than in the eastern part of the country. The species in the list are believed to be wholly wanting or of rare or local occurrence in that part of North America, which lies east and north of the meridian and parallel of Cleveland. Few of them are found in northern Ohio anywhere east of Erie county. The plants whose names are followed by an asterisk I have not found, but Mr. David F Day, of Buffalo, who collected at Toledo in 1865, tells me that he found them there.

Echinacea purpurea is inserted in the list because of a Toledo specimen in the herbarium of the Ohio State University.

Viola pedatifida.
Hypericum gymnanthum.
Hibiscus militaris.*
Aesculus glabra.

Polygala verticillata ambigua.
Desmodium lineatum.   '
Desmodium illinoense.
Petalostemon candidus.*

Silphium trifoliatum.
Solidago speciosa angustata.
Vernonia altissima.
Asclepias sullivantii.
Petalostemon violaceus.*
Psoralea melilotoides.
Geum vernum.
Pyrus angustifolia.
Spiræa lobata.
Ammannia coccinea.
Eryngium yuccæfolium.
Thaspium barbinode angustifolium
Valeriana pauciflora.
Actinella acaulis glabra.
Aster shortii.
Coreopsis aristosa.
Echinacea purpurea.
Eclipta alba.
Eupatorium altissimum
Helianthus grosse-serratus.
Helianthus hirsutus.
Helianthus mollis.
Helianthus occidentalis.
Helianthus parviflorus.
Helianthus tracheliifolius.
Kuhnia eupatorioides.
Liatris pycnostachya.*
Liatris squarrosa intermedia.
Prenanthes aspera.
Prenanthes crepidinea.
Rudbeckia triloba.

Phlox maculata.*
Hydrophyllum macrophyllum.
Phacelia purshii.
Cuscuta chlorocarpa.
Cuscuta decora.
Conobea multifida.
Gerardia auriculata.
Gratiola sphaerocarpa.
Seymeria macrophylla.
Tecoma radicans.
Lippia lanceolata.
Pycnanthemum muticum pilosum.
Scutellaria nervosa.
Scutellaria versicolor.
Euphorbia dentata.
Salix glaucophylla.
Iris cristata.
Smilax ccirrhata.
Trillium sessile.
Carex conjuncta.
Carex shortiana.
Carex stenolepis.
Carex granularis haleana.
Carex muhlenbergii enervis.
Cyperus refractus.
Rhynchospora cymosa.
Aristida gracilis.
Melica diffusa.
Poa brevifolia.
Triodia cuprea.
. Equisetum robustum.

A "List of Plants Observed Growing Wild in the Vicinity of Cincinnati," by C. G. Lloyd, with additions furnished by Walter H. Aiken, includes six hundred and forty-five species and varieties. Of these only fifty-one native species are lacking in Erie county. A greater number than this have been found in Lorain county, which borders Erie on the east, and might probably be found in each of the lake counties beyond.

DEFICIENCIES IN THE SANDUSKY FLORA.

Of the four counties, Lorain, Cuyahoga, Franklin and Licking, each two or three times as large as Erie, lists of plants have been published. Several hundred species are common to the four counties. Only four of these species, *Viola canadensis, Hieracium venosum, Veronica americana* and *Habenaria orbiculata*, have we failed to find in Erie county.

However twenty-five species not found in Erie county, grow in both Lorain and Cuyahoga. If we had complete lists for the counties farther east, Lake and Ashtabula, we should probably find in them a still larger number that do not grow in Erie county. Their higher hills and deeper ravines, give them a more northern flora, than one finds in the neighborhood of Sandusky. Moreover the Sandusky district contains no genuine bog or sphagnous swamp. Such a bog encircles a little lake a few miles south-east of Erie county in Camden township, Lorain county. The list of plants growing at Camden Lake and not in Erie county, is probably incomplete. For some of the names, I am indebted to Isabel S. Smith who has found the specimens in the Oberlin herbarium.

The list of other plants growing in northern Ohio is based mainly on the work of other collectors. It includes only those species which are said to grow in two or more counties bordering on the Lake. Of some of the species I have seen no specimens. Many other species have been reported and many others un-doubtedly grow in one place or another, but this list together with the catalogue of plants of Sandusky and vicinity and the plants of Camden are thought to include all the native phenogams and vascular cryp-togams which grow in the Lake counties, excepting such as are very rare or local.

PLANTS GROWING AT CAMDEN LAKE.

Coptis trifolia.
Sarracenia purpurea.
Nemopanthus fascicularis.
Potentilla palustris.
Viburnum cassinoides.
Cassandra calyculata.
Vaccinium oxycoccus.
Menyanthes trifoliata.
Alnus serrulata.
Arethusa bulbosa.
Habenaria orbiculata.

Pogonia ophioglossoides.
Smilacina trifolia.
Calla palustris.
Peltandra undulata.
Scheuchzeria palustris.
Carex canescens.
Carex debilis.
Carex trisperma.
Eriophorum virginicum album.
Rhynchospora alba.
Glyceria canadensis.

Woodwardia virginica.

OTHER PLANTS NOT FOUND NEAR SANDUSKY, BUT SAID TO GROW IN TWO OR MORE OF THE COUNTIES OF OHIO THAT BORDER ON LAKE ERIE.

Adlumia cirrhosa.
Corydalis glauca.
Stylophorum diphyllum.
Viola canadensis.
Viola hastata.
Viola rotundifolia.
Acer spicatum.
Polygala polygama.
Astragalus cooperi.
Prunus pennsylvanica.
Waldsteinia fragarioides.
Ribes oxyacanthoides.
Ribes rubrum subglandulosum.
Saxifraga virginiensis.
Oenothera biennis grandiflora.
Aralia hispida.
Diervilla trifida.
Lonicera ciliata.
Cornus canadensis.
Antennaria margaritacea.
Aster patens.
Cacalia suaveolens.

Hieracium venosum.
Polymnia uvedalia.
Solidago squarrosa.
Solidago uliginosa.
Pyrola secunda.
Rhododendron nudiflorum.
Vaccinium stamineum.
Monotropa hypopitys.
Phlox maculata.
Cynoglossum virginicum.
Melampyrum americanum.
Pentstemon laevigatus digitalis.
Veronica americana.
Rumex salicifolius.
Myrica asplenifolia.
Alnus incana.
Betula lutea.
Cypripedium parviflorum.
Pogonia verticillata.
Spiranthes latifolia..
Smilax glauca.
Uvularia perfoliata.

Veratrum viride.
Carex umbellata.
Cyperus erythrorhizos.
Milium effusum.

Larix americana.
Asplenium trichomanes.
Ophioglossum vulgatum.
Phegopteris polypodioides.
Woodsia obtusa.

EXTINCT SPECIES.

The only plant no longer found in the county, but known to have formerly grown in considerable quantity, is the Pitcher Plant, *Sarracenia purpurea* Mr. W. H. Todd remembers that this used to grow in the old huckleberry swamp near Axtell, in the eastern part of the county. This swamp of a hundred acres extent, is said to have produced yearly hundreds of bushels of blueberries, and a hundred bushels or so of cranberries. About 1856 a fire started in the muck, which lasted for a year, burning in places to a depth of four to six feet. This and drainage killed all the cranberries and nearly all the blueberries, and, how many other species, no body will ever know. It is now overgrown with a dense tangle of blackberry bushes interspersed with aspen and soft maple; the soil too light to be of much account. Had the original swamp been preserved, it would now be valuable for the berries it would produce. Only after repeated visits and prolonged searching in this wilderness by several persons, were two surviving bushes of the swamp blueberry discovered. Cranberries, which formerly grew also in a swamp near Berlin Heights, are now confined to a few square yards of ground, along a road near Milan.

Poison sumach formerly grew in the Axtell swamp. It is now all but extinct in the county. Leatherwood formerly abounded on Beecher's flats along the west branch of the Vermillion River. A single specimen remains, probably the only one in the county. A sedge collected on Cedar Point several years ago, and called by Prof. Wheeler, Cyperus Houghtonii, was afterward lost and so is not included in our catalogue. Likewise

we omit Strawberry Blite, *Chenopodium capitatum,*
seen on Green Island in 1892, but not collected, and
Hedeoma hispida, given in a list of plants, analyzed in
the eastern part of Erie county by Josephine Fish, a
number of years ago. The last has been found in
Lorain County by Prof. Kelsey, but perhaps is not indigenous to Ohio.

FOREST TREES.

Most of the land of Erie county is now under cultivation. Much of it was treeless when the earliest
settlements were made. Nevertheless, it supports a
greater variety of trees than do most of the counties of
Ohio, greater, perhaps, than any similar area farther
north in America. Birch, alder and tamarack, which
grow farther east in Ohio, are lacking in Erie county,
but it has ten kinds of oak, six of hickory, five of ash,
four of maple, four of poplar, four of willow, three of
thorn, two of elm, two of ironwood, two of wild crab,
and one each of black cherry, chokecherry, plum, june-
berry, basswood, box elder, buckeye, staghorn sumach,
papaw, tulip, cucumber, red-bud, locust, coffee-tree,
dogwood, pepperidge, sassafras, mulberry, hackberry,
buttonwood, beach, chestnut, walnut, butternut, hemlock, cedar and pine. Besides these, there are several
cultivated kinds that have become naturalized. The
distribution is given in the catalogue, where the names
may be found by referring to the index. Erie county
has five times as many native trees as the whole of
Great Britain.

THE CATALOGUE.

The catalogue that follows gives the names of the
phenogams and vascular cryptogams in the herbarium of the Sandusky high school which have been
collected in the region shown on the accompanying

map, i. e. Erie county, and the islands of Ottawa county, with the eastern portion of the peninsula, extending as far west as Port Clinton. Specimens of all the species and varieties have been examined by Prof. C. F. Wheeler, of Michigan, to whom I am indebted also for assistance in the determination of my earlier collections of Cyperaceæ and Naiadacæ, as well as of many puzzling forms found since.

Furthermore, a collection of most of the rarer species has been deposited in the Gray Herbarium, Cambridge, Mass., and another set in the Ohio State Herbarium, at the University at Columbus, and at both places botanists have examined them to see if there were errors in the identification.

To Dr. Erwin F Smith, of Washington, I am also indebted for valuable suggestions and assistance.

In a region where so many rare native species occur one would expect to find some exotic plants thriving, better than in most places in this latitude. As in the Philippine Islands where it has been introduced, so also in Sandusky, the tomato grows wild, coming up like a weed in many places, but especially along the bay shore, where it ripens its fruit year after year. It is difficult in some cases to say whether a species is naturalized or not. Oats grow on the shores of the islands, as well as about the docks in Sandusky, and along roads, but herbs of which all the specimens found have probably sprung directly from the seeds of cultivated plants, are not included in the catalogue. A watermelon vine with fruit was found on the shore of Cedar Point, and this and muskmelon, squash and pumpkin, on waste ground in Sandusky near the Bay. Peanuts, which are raised in small quantities by many people in and near Sandusky, have been found spontaneous in two places in the city. Snapdragon, gilliflower, candytuft, common honesty, petunia, and others, have been found growing in waste places, but are excluded from the catalogue under the rule given above. On the other hand plants that are never cultivated in this region are included, even if merely adventive.

In nomenclature I have, in the main, followed the Index Kewensis, giving in parenthesis the names used in the sixth edition of Gray's Manual, in the few cases where those differ materially from the names of the Kew Index. Names of species not native to this part of the world, are printed in italics. An asterisk indicates that the species is at present known to grow in few, if any places in Ohio, except in the neighborhood of Sandusky.

Relative abundance is expressed by the following terms in the order named; rare, scarce, infrequent, frequent, common, abundant. When standing alone or coming first, they refer to Erie county as a whole.

CATALOGUE.

PTERIDOPHYTA.

OPHIOGLOSSACEÆ.

BOTRYCHIUM, Swartz.

B. TERNATUM, Swartz.
Eastern Milan, Berlin, Florence, Vermillion; infrequent. Varies greatly.

B. VIRGINIANUM, Swartz.
Frequent. Put-in-Bay.

FILICES.

ADIANTUM, L.

A. PEDATUM, L. Maiden-hair Fern.
Common. Not on the Islands.

ASPIDIUM, Swartz.

A. ACROSTICHOIDES, Swartz. Shield Fern.
Scarce in Perkins. Common on high banks of Huron and Vermillion Rivers.

A. CRISTATUM, Swartz.
Vermillion River bottoms, Florence; rare.

A. GOLDIANUM, Hook.
Florence and Kromer's woods, Perkins; scarce.

A. MARGINALE, Swartz.
Common on steep river banks.

A. NOVEBORACENSE, Swartz.
Infrequent.

A. SPINULOSUM, Swartz.

Frequent in rich woods.

A. SPINULOSUM INTERMEDIUM, D. C. Eaton.

Frequent. Neither this nor the species seen on Peninsula or Islands.

A. THELYPTERIS, Swartz.

Common.

ASPLENIUM, L. Spleenwort.

A. ANGUSTIFOLIUM, Michx.

Infrequent.

A. EBENEUM.

Common in Furnace woods, Vermillion, "Cedar Point," J. R. Schacht.

A. FILIX-FOEMINA, Bernh.

Common. Not on Peninsula or Islands.

A. THELYPTEROIDES, Michx.

Perkins and Florence ; scarce.

CAMPTOSORUS, Link, Walking-fern.

C. RHIZOPHYLLUS, Link.

On sides of sandstone rocks, Vermillion River, S. Florence ; on limestone, three places in Margaretta, Catawba, Kelley's sland.

CYSTOPTERIS, Bernh, Bladder Fern.

C. BULBIFERA, Bernh.

Frequent. Islands.

C. FRAGILIS, Bernh.

Common. Kelley's Island.

DICKSONIA, L'Her.

D. PILOSIUSCULA, Willd.

Vermillion River ; frequent. Big woods, Perkins ; scarce.

ONOCLEA, L.

O. SENSIBILIS L. Sensitive Fern.

Common. Not on the Islands.

O. STRUTHIOPTERIS, Hoffman.
Vermillion River bottoms, frequent.

OSMUNDA, L.

O. CINNAMOMEA, L. Cinnamon Fern.
Infrequent; Florence, Milan "Perkins."
O. CLAYTONIANA, L.
Common in moist woods. Not on Peninsula or
Islands.
O. REGALIS, L. Flowering Fern.
Infrequent in wet woods.

PELLÆA, Link, Cliff-Brake.

P. ATROPURPUREA, Link.
Sandstone quarry, Furnace woods, Vermillion;
on limestone, Margaretta, Peninsula, Catawba,
Kelley's Island, Put-in-Bay.

PHEGOPTERIS, Fee, Beech Fern.

P. HEXAGONOPTERA, Fee.
Frequent from the Huron river east.

POLYPODIUM, L., Polypody.

P. VULGARE, L.
Rocky banks of rivers and Kelley's Island;
scarce.

PTERIS, L.

P. AQUILINA, L., Common Brake.
Frequent.

EQUISETACEÆ.

EQUISETUM, L., Horsetail.

E. ARVENSE, L.
Common but not observed on the Islands, except
Kelley's.

E. LAEVIGATUM, Braun.

Frequent, at least in the western part of the county.

E. LIMOSUM, L.

Lake marshes, Huron Tp.

E. LITTORALE, Kuhl.*

Perkins; rare.

E. PRATENSE, Ehrh.

Frequent.

E. ROBUSTUM, Braun,

Common, apparently entirely supplanting E. *hyemale.* Put-in-Bay and Kelley's Island but no others

E. VARIEGATUM, Schleicher,*

Cedar Point and elsewhere; rare.

LYCOPODIACEÆ.

LYCOPODIUM, L., Club-Moss.

L. COMPLANATUM, L. Ground-Pine.

East fork of Vermillion River; rare.

L. DENDROIDEUM, Michx.

East of Milan; rare.

L. LUCIDULUM, Michx.

Quarry in Furnace woods, Vermillion; rare.

Each of the three kinds of club-moss has been found in but a single spot, and of the last two, only a few specimens.

GYMNOSPERMÆ.

CONIFERÆ.

JUNIPERUS, L.

J. COMMUNIS, L.

Mr. Latham's woods, Catawba; very rare.

J. VIRGINIANA, L., Red Cedar.

Frequent in dry soil in various parts of Erie and Ottawa counties. Formerly abundant on the islands where its wood was one of the first sources of income to the early settlers. Many stumps two feet or more in diameter still remain on Kelley's Island, though they are being used for kindling and for boat knees. The trees grew in the thin soil overlying the limestone, and so the roots following the level surface of the rock were given off from the trunk at a right angle. Having greater strength than an artificial joint and great durability sections of these stumps make excellent knees for small boats. Large cedars grew formerly also on Cedar Point where small ones are common now.

PINUS, Tourn.

P. STROBUS, L. White Pine.

Cedar Point and Vermillion River. Both this and Red Cedar grew once where Sandusky Bay is now.

TAXUS, Tourn.

T. CANADENSIS, Willd. American Yew. Ground Hemlock.

Shores of Islands and Vermillion River; infrequent.

TSUGA, Carriere.

T. CANADENSIS, Carr. Hemlock.

Common along the Old Woman Creek at Berlin Heights and along the Vermillion River.

MONOCOTYLEDONES.

TYPHACEÆ.

SPARGANIUM, Tourn, Bur-reed.

S. ANDROCLADUM. Engelm.
Lake marshes.   Middle Bass.

S. EURYCARPUM, Engelm.
Lake marshes.   Middle Bass.

S. SIMPLEX, Huds.
Southern Florence, Shinrock.

TYPHA, Tourn.

T. AUGUSTIFOLIA, L.
Castalia stream, Portage River and North Bass;
scarce.

T. LATIFOLIA, L.   Common Cat-tail.
Common.

NAIADACEAE.

NAIAS, L., Naiad.

N. FLEXILIS, Rostk, and Schmidt.
Common.

N. FLEXILIS ROBUSTA, Morong.*
Infrequent.

N. GRACILLIMA, A. Br.*
‹   "Portage River" A. J. Pieters.

POTAMOGETON, Tourn.   Pond-weed.

P. AMPLIFOLIUS, Tuckerm. Deep water; infrequent.

P. FOLIOSUS, Raf.
East Harbor, Put-in-bay, North Bass; mostly in
shallow water.

P. FOLIOSUS NIAGARENSIS, (Tuckerm.)   Morong.*
North Bass and small streams in Erie County,
especially Mills Creek.

P. FRIESII, Rupr.*
Sandusky Bay, Put-in-Bay; infrequent.

P. HETEROPHYLLUS, Schreb.*
Frequent; especially the variety *longipeduncula-tus.* The variety *maximus* occurs at North Bass.

P. HILLII, Morong.*
East Harbor; rare.

P. INTERRUPTUS, Kitaibel.*
Sandusky Bay, Put-in-Bay; rare.

P. LONCHITES, Tuckerm.
Common.

P. LUCENS, L.*
Frequent.

P. NATANS, L.
Common, as is also the so called variety, *prolixus.*

P. PECTINATUS, L.
Abundant-

P. PERFOLIATUS, L.
Frequent.

P. PERFOLIATUS RICHARDSONII A. Bennett.
Abundant.

P. PRAELONGUS, Wulf.*
Sandusky Bay, August Guenther. Perhaps its habit of withdrawing beneath the water, as soon as its fruit is set, has prevented us from finding much of it.

P. PUSILLUS, L.*
Infrequent.

P. ROBBINSII, Oakes.
Sandusky Bay; scarce.

P. ZIZII, Roth.*
Sandusky Bay; scarce.

P ZOSTERÆFOLIUS, Schum.
Common.

TRIGLOCHIN, L. Arrow-Grass.

T. PALUSTRE, L.*
Castalia Sporting Club grounds; rare.

ZANNICHELLIA, Mitchell,   Horned Pond-Weed.

Z. PALUSTRIS.

The "variety" *pedunculata* grows. or did grow
in one of the rivulets flowing from the Blue Hole,
Castalia; rare.

ALISMACEÆ.

ALISMA, L.   Water-Plantain.

A. PLANTAGO, L.

Common.   ·

LOPHOTOCARPUS, T. Durand.

L. CALYCINUS, (Engelm) J. G. Smith.*

In   a   small   pond   bordering   the   southern
boundary of Sandusky.

SAGITTARIA, L.   Arrow-Head.

S. ARIFOLIA, Nutt.

Oxford, Danbury ; scarce.

S. GRAMINEA, Michx.*

Sandusky Bay. "East Harbor," A. J. Pieters.

S. LATIFOLIA, Willd. (S. VARIABILIS, Engelm.)

Common and variable.

S. RIGIDA, Pursh. (S. HETEROPHYLLA, Pursh.)

Sandusky Bay, Put-in-Bay, Harbors ; frequent.
In deeper water than the last.

HYDROCHARIDACEÆ.

ELODEA, Michx.   Water-Weed.

E. CANADENSIS, Michx.

Common. Kelley's Island, Put-in-Bay. Filling
the cove east of Sandusky so as to make it
difficult to row a boat there.

VALLISNERIA, L. Tape-Grass, Eel-Grass.

V. SPIRALIS, L.
Common.

GRAMINEÆ.

AGROPYRON, Gaert.

A. CANINUM, Beauv.*
Berlin Heights; rare. .
A. GLAUCUM, R. & S.*
L. S. & M. S. Ry., Sandusky; scarce.
A. REPENS, Beauv. Couch-Grass, Quitch-Grass.
Infrequent. Kelley's Island.

AGROSTIS, L. Bent-Grass.

A. ALBA, L.
Common, as is the variety *vulgaris*, Red Top.
A. PERENNANS, Tuckerm, Thin-Grass.
Frequent.
A. SCABRA, Willd. Hair-Grass.
Infrequent. Put-in.Bay, Middle Bass.

ALOPECURUS, L. Foxtail-Grass.

A. GENICULATUS ARISTULATUS, Torr.
Islands, Peninsula and Milan; rare in Erie county.

AMMOPHILA, Host.

A. ARUNDINACEA, Host. Sea Sand-Reed.
Cedar Point and Marblehead Sand Spit.

ANDROPOGON, L. Beard-Grass.

A. PROVINCIALIS, Lam. (A. FURCATUS, Muhl.)
Frequent.
A. SCOPARIUS, Michx.
Frequent. Not observed in Ottawa county.

ARISTIDA, L. Triple-awned Grass.

A. GRACILIS, Ell.*
> Unplowed prairle, Perkins.

A. PURPURASCENS, Poir.*
> Roadside, Joseph Smith's, Perkins.

ASPERELLA, Humb. Bottle-brush Grass.

A. HYSTRIX, Humb·
> Common.

BOUTELOUA, Lag. Muskit-Grass.

B. RACEMOSA, Lag.
> Castalia cemetery and southwest. Marblehead ;
> dry ground ; scarce.
> Our forms approach the variety *aristosa.*

BRACHYELYTRUM, Beanv.

B. ERECTUM, Beanv. (B. ARISTATUM R. & S.)
> Frequent.

BROMUS, L. Brome-Grass.

B. CILIATUS, L.
> Common. Kelley's Island., Rattlesnake Island.
> The variety *purgans* also common, but not on
> the Islands.

B. KALMII, Gray. Wild Chess.
> Margaretta Ridge ; rare.

B. *racemosus*, L. Upright Chess.
> Common.

B. *secalinus*, L Cheat or Chess.
> Not so common as the last.

B. *tectorum*, L.
> Along Big Four Ry., Sandusky and Castalia ;
> elsewhere also, but scarce.

CENCHRUS, L. Hedgehog or Bur-Grass.

C. TRIBULOIDES, L.
> Common in sand.

CHRYSOPOGON, Trin.

C. NUTANS, Benth, Indian Grass, Wood Grass.
Frequent.

CINNA, L. Wood Reed-Grass.

C. ARUNDINACEA, L.
Frequent.

DACTYLIS, L. Orchard-Grass.

D. *glomerata*, L.
Frequent.

DANTHONIA, D C. Wild Oat-Grass.

D. SPICATA, A. & S.
Common. Not on Islands, except Put-in-Bay.

DEYEUXIA, Raf.

D. CANADENSIS, Beauv. Blue-Joint Grass.
Frequent. Middle Bass, North Bass.

EATONIA, Raf.

E. OBTUSATA, Gray.*
Infrequent. Margaretta Idge, Marblehead,
North Bass, etc.

E. PENNSYLVANICA, Gray.
Frequent. Islands.

E. PURPURASCENS Raf. (E. DUDLEYI, Vasey. E. NITIDA
Nash.)
Florence. and Furnace woods, Vermillion.

ELEUSINE, Gaertn.

E. *indica*, Gaertn. Dog's-tail or Wire Grass.
Formerly seldom seen, but now common along
many sandy lanes.

ELYMUS, L. Lyme-Grass, Wild Rye.

E. CANADENSIS, L.
Frequent, especially on sand beaches. Islands.
The so called variety *glaucifolius* occurs in a
number of places but does not appear at all distinct.

E. STRIATUS, Willd.

> Infrequent. Kelley's Island. The so called variety *villosus* was found in Perkins.

E. VIRGINICUS, L

Frequent along streams and shores of the Islands.

ERAGROSTIS, Host.

E. CAPILLARIS, Nees.

> Willow Point, Margaretta and different parts of the Peninsula.

E. FRANKII, Steud.

> Perkins, Castalia, Lockwood's woods, Catawba.

E. *major*, Host.

> Abundant.

E. PURSHII, Schrader.

> Common in Erie Co., especially along railroads. Kelley's Island.

E. REPTANS, Nees.

> Infrequent.

E. SPECTABILIS, Steud.* (E. PECTINACEA SPECTABILIS, Gray) Lake sands of Cedar Point, Marble-head Spit, and Port Clinton; local.

FESTUCA, L. Fescue-Grass.

F *elatior*, L. Meadow Fescue.

> Common. The variety *pratensis* common in Sandusky and along some country roads.

F. NUTANS, Spreng.

> Common. Not noticed on Kelley's sland.

F. TENELLA, Willd.

> Marblehead, Cedar Point and east of Milan.

GLYCERIA, R. Br. Manna Grass.

G. FLUITANS, R. Br.

> Infrequent. Islands.

G. NERVATA, Trin.

> Common.

G. PALLIDA, Trin.
> Port Clinton; rare.

HIEROCHLOE, S. G. Gmel.

H. BOREALIS, R. & S.
> "Perkins" Elon House.

HORDEUM, L.

H. JUBATUM, L.   Squirrel-tail Grass.
> Common along L. S. & M. S. Ry. in Ottawa Co.
> Blue Hole, Castalia.   Kelley's Island, where
> probably introduced in baled hay.   Marblehead.

KOELERIA, Pers.

K. CRISTATA, Pers.*
> Catawba, where first found by A. D. Selby.
> Margaretta Ridge, Oxford; also ten miles west of
> Toledo.

LEERSIA, Swartz.

L. ORYZOIDES, Swartz.   Rice Cut-grass.
> Common.

L. VIRGINICA, Willd.   White Grass.
> Common but not noticed on any island except
> Kelley's.

LOLIUM, L.

L. *perenne*, L.   Common Darnel, Ray or Rye-Grass.
> Sandusky, Soldier's Home, Kelley's Island, Put-in-
> Bay; infrequent.   Not noticed until 1897.

MELICA, L.   Melic-Grass.

M. DIFFUSA, Pursh.*
> Castalia; rare.

MUHLENBERGIA, Schreb.   Drop-seed Grass.

M. GLOMERATA, Trin.*
> West of Castalia; rare; also ten miles west of
> Toledo.

M. MEXICANA, Trin.
  Common.

M. SCHREBERI, J. F. Gmel. (M. DIFFUSA, Schreb.)
  Common.

M. SOBOLIFERA, Trin.*
  Florence, Catawba; rare.

M. SYLVATICA, Torr. & Gray.
  Perkins, Florence, Middle Bass; infrequent.

M. WILLDENOWII, Trin.
  Vermillion River, Huron, Milan, Perkins, Margaretta Ridge; infrequent.

ORYZOPSIS, Michx. Mountain Rice.

O. MELANOCARPA, Muhl.
  Margaretta Ridge, Vermillion River, Put-in-Bay; rare.

PANICUM, L. Panic-Grass.

P. AGROSTOIDES, Muhl.
  Huron, Milan, Oxford, Perkins, North Bass; local.

P. BARBULATUM, Michx.
  Berlin; rare.

P. CAPILLARE, L. Old-witch Grass.
  Common.

P. CLANDESTINUM, L.
  Cedar Point, Perkins, and common along river channels.

P. COLUMBIANUM, Scribn.
  Castalia, Cedar Point. Formerly called P. *dichotomum*.

P. *crus-galli*, L. Barnyard-Grass.
  Abundant.

P CRUS-GALLI HISPIDUM, Muhl.
  Frequent on wet ground about Sandusky Bay and East Harbor.

P. DEPAUPERATUM, Muhl.
  Catawba and high banks of Vermillion River and Old Woman Creek.

P. DICHOTOMUM, L.

Common and variable, the so called variety *gracile*, found only at Berlin Heights, seeming most distinct from other forms.

P. FLEXILE, Scribn.

Castalia prairie; common. Oxford.

P. *glabrum*, Gaudin. Small Crab-Grass.

Common. North Bass the only island.

P. LATIFOLIUM, L.

Common in Erie County.

P. MILIACEUM, L. Millet.

Adventive. "Cedar Point," E. Claassen. Sandusky near the Bay, one specimen growing on rubbish.

P. PROLIFERUM, Lam.

Sandusky, Oxford; rare.

P. PUBESCENS, Lam.

Common in Erie County.

P. *sanguinale*, L. Large Crab-Grass.

Abundant.

P. SCOPARIUM, Lam.

Oxford, Margaretta, Cedar Point, Port Clinton; common.

P. VIRGATUM, L.

Frequent. Kelley's Island. Abundant on sandy shores of Lake Erie.

PASPALUM, L.

P. SETACEUM, Michx.

Dell Lindsley's orchard. Perkins, where it has probably been for many years.

PHALARIS, L.

P. ARUNDINACEA, L.

Cedar Point, Huron, western Margaretta, Middle Bass; infrequent. The variety *picta* Ribbon-Grass, has become established along some road-side ditches.

P. *canariensis*, L. Canary-Grass.

Adventive in Sandusky.

PHLEUM, L.

P *pratense*, L. Timothy.
  Abundant.

PHRAGMITES, Trin. Reed.

P. COMMUNIS, Trin.
  Frequent on wet ground. Huron, Castalia, Port
  Clinton, Harbors.

POA, L. Meadow-Grass.

P. ALSODES, Gray.
  Florence; scarce.
P. *annual*, L. Low Spear-Grass.
  Frequent.
P. *compressa*, L. Wire-Grass.
  Abundant.
P. DEBILIS, Torr.
  Furnace woods, Vermillion; rare.
P. PRATENSIS, L. June Grass. Kentucky Blue-Grass.
  Abundant. One specimen has a panicle eleven
  inches long.
P. SEROTINA, Ehrhart.
  Huron; rare.
P. SYLVESTRIS, Gray.
  Parker's Creek, Florence; rare.
P. *trivialis*, L.
  Shinrock; rare.

SETARIA, Beanv.

S. *glauca*, Beanv. Foxtail. Pigeon-Grass.
  Abundant. The worst weed we have.
S. *italica*, Beanv. Italian Millet, Hungarian Grass.
  Rarely escaped. Middle Bass, North Bass.
S. *verticillata*, Beanv.
  Sandusky near Big Four dock, 1898.
S. *viridis*, Beanv. Green Foxtail.
  Less abundant than S. *glauca*.

SPARTINA, Schreb. Marsh Grass.

S. SCHREBERI, J. F. Gmel. (S. CYNOSUROIDES Willd.)
Fresh-water Cord-Grass.
Frequent. Middle Bass.

SPOROBOLUS, R. Br. Rush-Grass.

S. ASPER, Kunth.
L. S. & M. S. Ry, east of Sandusky ; rare.
S. CRYPTANDRUS, Gray.
Frequent on Cedar Point and several places on
the Peninsula.
S. NEGLECTUS, Nash.
Sandusky, Castalia, Plaster Beds.
S. VAGINÆFLORUS, Vasey.
Common. Kelley's and Put-in-Bay the only
Islands.

STIPA, L.

S. SPARTEA, Trin.* Porcupine Grass.
In sand; Cedar Point, Perkins, Bloomingville
cemetery ; rare.

TRIODIA, R. Br.

T. CUPREA, Jacq.* Tall Red-Top.
In sand near the road through the woods between Port Clinton and Catawba; rare.

TRIPLASIS, Beauv.

T. PURPUREA, Chapm. Sand-Grass.
Frequent on all sandy shores of Lake Erie; in
places abundant. Kelley's Island.

ZIZANIA, L.

Z. AQUATICA, L. Indian Rice. Water Oats.
Abundant in shallow parts of Sandusky Bay, the
Harbors, etc. Middle Bass.

CYPERACEÆ.

CAREX, L., Sedge.

C. ALBICANS, Willd.*
Put-in-Bay; rare.

C. ALBURSINA, Sheldon, (C. LAXIFLORA LATIFOLIA Boott.)
Frequent. Kelley's Island.

C. AQUATILIS, Wahl.
Huron, Cedar Point, Put-in-Bay; scarce.

C. ARCTATA, Boott.
Florence, Berlin, Oxford; rare.

C. AUREA, Nutt.*
One vigorous plant growing on a stump that stands in a stream near the Blue Hole, Castalia.

C. BICKNELLII, Britton (C. STRAMINEA CRAWEI, Boott.)
Berlin Heights; rare.

C. BROMOIDES, Schkuhr.
Florence, Berlin Heights, Milan; local.

C. CAREYANA, Torr.
Beecher's flats, Vermillion River; rare.

C. CEPHALOIDEA, Dewey.
Frequent.

C. CEPHALOPHORA, Muhl.
More frequent than the last. Bass Islands.

C. COMMUNIS, Bailey.
Florence, Margaretta Ridge; scarce.

C. COMMUNIS WHEELERI, Bailey.*
Vermillion River, Florence; rare.

C. CONJUNCTA, Boott.
Florence, Berlin, Perkins; scarce.

C. CRAWEI, Dewey.*
Castalia prairie, Marblehead; local.

C. CRINITA, Lam.
Frequent from the Huron River east, especially in Berlin.

C. DAVISII, Schwein & Torr.

Shinrock, Perkins, Port Clinton, Kelley's Island; infrequent.

C. DIGITALIS, Willd.

Common in Florence; frequent in Vermillion, Berlin and Milan.

C. DIGITALIS COPULATA, Bailey.

Florence, Berlin, Milan; frequent.

C. EBURNEA, Boott.*

Kelley's Island. Put-in-Bay; rare.

C. FILIFORMIS, L.

Frequent?

C. FŒNEA PERPLEXA, Bailey.*

Furnace woods, Vermillion; rare.

C. FUSCA, All.

Throughout Erie Co; infrequent.

C. GLAUCODEA, Tuckerm.

Vermillion, Berlin, Milan; infrequent.

C. GRACILLIMA, Schwein.

Frequent in Erie Co.

C. GRANULARIS, Muhl.

Frequent. Kelley's Island.

C. GRANULARIS HALEANA, Porter.* (C. HALEANA, Olney)

Florence, Castalia, Groton; infrequent.
The Groton specimens have very broad leaves.

C. GRAYII, Carey.

Huron, Milan and east; infrequent.

C. GRISEA, Wahl.

Rather frequent.

C. HITCHCOCKIANA, Dewey.

Florence; scarce.

C. HYSTERICINA, Muhl.

Common. Put-in-Bay and Middle Bass the only islands.

C. INTERIOR, Bailey.*

Castalia; rare.

C. INTUMESCENS, Rudge.

Berlin, Vermillion, Florence; infrequent.

C. JAMESII, Schwein.
>Berlin; rare. Florence; infrequent.

C. LANUGINOSA, Michx. (C. FILIFORMIS LATIFOLIA, Boeckl.)
>Frequent. Put-in-Bay.

C. LAXICULMIS, Schwein.
>Florence, Vermillion, Milan; infrequent.

C. LAXIFLORA, Lam.
>Frequent. Kelleys Island.

C. LAXIFLORA PATULIFOLIA, Carey.
>Florence, Berlin, Huron; infrequent.

C. LAXIFLORA STRIATULA, Carey.
>Common.

C. LAXIFLORA VARIANS, Bailey.
>Common. Kelley's the only island.

C. LUPULINA, Muhl.
>Common. The so-called variety *hedunculata* occurs in Florence.

C. LURIDA, Wahl.
>Frequent. Hybrids of this and the last occur in Florence and Berlin.

C. MONILE, Tuckerm.
>Vermillion, Berlin, Kimball; scarce.

C. MUHLENBERGII, Schkuhr.*
>Cedar Point; frequent.

C. MUHLENBERGII ENERVIS, Boott.
>Catawba; rare.

C. *muricata*, L.
>Furnace woods, Vermillion; rare.

C. MUSKINGUMENSIS, Schwein.*
>Catawba; rare.

C. OLIGOCARPA, Schkuhr.
>Prout's, Shinrock, Vermillion, Florence: infrequent.

C. PALLESCENS.
>Berlin Heights and Florence; rare.

C. PEDUNCULATA, Muhl.
>Steep banks of Vermillion River, Florence; rare.

C. PENNSYLVANICA, Lam.
>Abundant. Put-in-Bay the only island.

C. PLANTAGINEA, Lam.
>Steep banks of Vermillion River and tributaries in southern Florence; infrequent.

C. PLATYPHYLLA, Carey.
>High banks, Vermillion River, Florence; rare.

C. PRASINA, Wahl.
>Infrequent.

C. PSEUDO-CYPERUS AMERICANA, Hochst. (C. COMOSA, Boott.)
>Islands, Cedar Point, Castalia, South Florence; local.

C. PUBESCENS, Muhl.
>Frequent, especially in Florence.

C. RICHARDSONII, R. Br.*
>Castalia cemetery; rare.

C. RIPARIA, Curtis.
>Infrequent.

C. ROSEA, Schkuhr.
>Common. Middle Bass the only island.

C. ROSEA RADIATA, Dewey.
>Florence; rare.

C. SARTWELLIANA, Olney.* (C. SARTWELLII, Dewey.)
>Castalia; scarce. Huron, Cedar Point; rare.

C. SCABRATA, Schwein.
>Springy banks of Vermillion River; rare.

C. SCOPARIA, Schkuhr.
>Common. Not on the Islands.

C. SETACEA, Dewey.
>Oxford; rare or else taken for C. *vulpinoidea*.

C. SHORTIANA, Dewey.
>Perkins, Castalia, Berlin and common in Milan.

C. SICCATA, Dewey.*
>Perrin's, Milan; Margaretta Ridge; rare.

C. SPARGANIOIDES, Muhl.
>Frequent. Kelley's Island. Rattlesnake.

C. SQUARROSA, L.
>   Frequent.

C. STENOLEPIS, Torr.
>   Common, especially near Sandusky. Middle
>   Bass the only island.

C. STERILIS, Schkuhr.
>   Castalia; rare.

C. STERILIS CEPHALANTHA, Bailey.
>   Tisdell's, Vermillion; rare.

C. STIPATA, Muhl.
>   Common.

C. STRAMINEA, Willd.
>   Infrequent.

C. STRAMINEA BREVIOR, Dewey.*   (C. FESTUCACEA,
>   Schkuhr.)
>   Marblehead, Johnson's Island, Kelley's, Green.

C. STRAMINEA MIRABILIS. Tuckerm.
>   Huron, Milan and east; rare.

C. STRICTA, Lam.
>   Scarce.

C. STRICTA DECORA, Bailey.*   (C. HAYDENII, Dewey.)
>   Kimball; rare.

C. TENELLA, Schkuhr.
>   Vermillion River flat, Florence; one place.

C. TERETIUSCULA, Gooden.
>   Castalia; scarce.

C. TETANICA, Schkuhr.
>   Castalia prairie; frequent.

C. TETANICA MEADII, Bailey.*
>   Castalia prairie.

C. TETANICA WOODII, Bailey.
>   Huron and southern Florence in woods.   Differs
>   from the species in habitat and appearance.

C. TORTA, Boott.
>   nfrequent.   One specimen considered a hybrid
>   of this and C. *crinita*.

C. TRIBULOIDES, Wahl.
> Frequent, especially the variety *turbata.*   North
> Bass.

C. TRIBULOIDES CRISTATA, Bailey.
> Common.   North Bass the only island.

C. TRIBULOIDES REDUCTA, Bailey.*
> Florence and Huron ; rare.

C. TRICEPS HIRSUTA, Bailey.
> Frequent.

C. TRICHOCARPA, Muhl.
> Huron River, Milan.   The variety *imberbis*
> grows in Florence.   Both scarce.

C. TRICHOCARPA ARISTATA, Bailey.*
> Huron, Castalia ; infrequent.

C. TVPHINOIDES, Schwein.
> East Berlin ; local.

C. TUCKERMANNI, Boott.
> Infrequent.

C. UTRICULATA, Boott.
> Blair Creek, Florence ; rare.   The so-called
> variety *minor* at Tisdell's, Vermillion ; rare.

C. VARIA, Muhl.
> Frequent.

C. VIRESCENS, Muhl.
> Oxford, Huron and east; common.

C. VIRESCENS COSTATA, Dewey.
> Berlin Heights and east; infrequent.

C. VIRIDULA, Michx.* (C. FLAVA VIRIDULA, Bailey.)
> Castalia prairie; local.

C. VULPINOIDEA, Michx.
> Common.

C. WILDENOWII, Schkuhr.
> Florence; rare.

CLADIUM, P. Br. Twig-Rush.

C. TRIGLOMERATUM, Nees. (C. MARISCOIDES, Torr.)
> Perkins, "Castalia," E. Claassen.

CYPERUS, L. Galingale.

C. DIANDRUS, Torr.
Frequent. Islands. The so-called variety *castaneus* on Cedar Polnt.

C. ESCULENTUS, L.
Frequent in cultivated ground.

C. FILICULMIS, Vahl, (MARISCUS GLOMERATUS, Barton.)
Rather frequent in sand.

C. FLAVESCENS, L.*
Frankinberg's pasture, south-eastern Florence.

C. MICHAUXIANUS, Schult, (C. SPECIOSUS, Vahl.)
About Sandusky Bay ; scarce.

C. REFRACTUS, Engelm.*
East branch, Vermillion River ; one specimen.

C. SCHWEINITZII, Torr.
Cedar Point; common. Port Clinton.

C. STRIGOSUS, L.
Common and variable. Abundant in many pastures. One specimen over three feet tall has primary rays 8 inches long, secondary rays 2½ inches, spikelets nearly 1 inch.

DULICHIUM, Pers.

D. SPATHACEUM, Pers.
Perkins, Milan, Cedar Point; local.

ELEOCHARIS, R. Br. Spike-Rush.

E. ACICULARIS, R. Br.
Castalia and borders of marshes connected with Lake Erie; frequent. Bass Islands.

E. ACUMINATA, Nees.* (E. COMPRESSA, Sullivant.)
Sandusky, Cedar Point, Huron, Marblehead; scarce.

E. ENGELMANNI, Steud.*
North of Tisdell's, Vermillion ; rare.

E. INTERMEDIA, Schult.
Cedar Point, Johnson's Island, Marblehead, Bass Islands; frequent.

E. OVATA, R. Br.
Frequent. Kelley's Island. North Bass.

E. PALUSTRIS, R. Br.
Frequent.

E. PALUSTRIS GLAUCESCENS, Gray.
Frequent. Put-in-Bay.

E. PALUSTRIS VIGENS, Bailey.*
Sandusky Bay; in water several feet deep.

E. TENUIS. Schult.
Infrequent.

ERIOPHORUM, L. Cotton-Grass.

E. POLYSTACHYON, L.
"Huron River" Henry Schoepfle.

FIMBRISTYLIS, Vahl.

F. AUTUMNALIS, R. & S.
A little bog near the Cedar Point light house.

F. CAPILLARIS, Gray.*
In sand, south Perkins and east of Milan ; local.

RYNCHOSPORA, Vahl. Beak-Rush.

R. CAPILLACEA, Torr.
Prairie along L. E. & W. Ry., west of Castalia ;
local.

R. CYMOSA, Nutt.*
East of Milan ; local.

R. GLOMERATA, Vahl.*
East of Milan; local. Also ten miles west of
Toledo.

SCIRPUS, L. Bulrush.   .

S. ATROVIRENS.
Common.

S. ERECTUS, Poir.* (S. DEBILIS, Pursh.)
Along shore of East Harbor west of Lakeside.

S. ERIOPHORUM, Michx. (ERIOPHORUM CYPERINUM, L.)
Frequent. The variety *laxum* occurs in Florence,
Milan, and, probably, elsewhere.

S. LACUSTRIS, L. Great Bulrush.
Common. Extensively used in the vineyards for
tying up the vines.

S. LINEATUS, Michx. (ERIOPHORUM LINEATUM, Benth &
Hook.)
Frequent. Kelley's Island. North Bass.

S. MARITIMUS, L. (S. FLUVIATILIS, Gray.) River Club-
Rush.
Common in the marshes east of Sandusky and in
the East Harbor; elsewhere infrequent. Put-in-
Bay.

S. POLYPHYLLUS, Vahl.
Frequent. Middle Bass.

S. PUNGENS, Vahl.
Common, especially about Sandusky Bay and
Lake Erie.

S. SYLVATICUS, L.*
"Pond near U. S. Fish Hatchery, Put-in-Bay."
A. J. Pieters.

S. TORREYI, Olney.*
North side of East Harbor; rare.

SCLERIA, Berg. Nut-Rush.

S. PAUCIFLORA, Muhl.*
East of Milan; local. Also ten miles west of
Toledo.

S. TRIGLOMERATA, Michx.*'
East of Milan; local. Also ten miles west of
Toledo.

ARACEÆ.

ACORUS, L. Sweet Flag.

A. CALAMUS, L.
Frequent. Abundant near Port Clinton. Put-in-
Bay. "Kelley's Island."

ARISÆMA, Mart.

A. DRACONTIUM, Schott. Green Dragon, Dragon-root.
Scarce.

A. TRIPHYLLUM, Schott. Indian Turnip.
Common.

SYMPLOCARPUS, Salisb. Skunk Cabbage.

S. FOETIDUS, Nutt.
Infrequent.

LEMNACEÆ.

LEMNA, L. Duck-weed, Duck's-meat.

L. MINOR, L.
Common at Castalia and on still water connected
with Lake Erie. Islands.

L. POLYRRHIZA, L. (SPIRODELA POLYRRHIZA, Schleid.)
Common on still water connected with the Lake.
Florence.

L. TRISULCA, L.
Castalia and still waters connected with the
Lake; infrequent- Put-in-Bay.

WOLFFIA, Workel.

W. COLUMBIANA, Karsten.
Mouth of Old Woman Creek, Pipe Creek, Put-in-
Bay; local.

COMMELINACEÆ.

TRADESCANTIA, L. Spiderwort.

T. VIRGINIANA, L.
Frequent, especially on Cedar Point.

T. VIRGINIANA OCCIDENTALIS, Britton.
B. & O. Ry. seven miles south of depot; rare.

PONTEDERIACEÆ.

HETERANTHERA, Ruiz & Pav. Mud-Plantain.

H. GRAMINEA, Vahl.
Common in still water connected with Lake Erie.

PONTEDERIA, L. Pickerel-weed.

P. CORDATA, L.
Frequent in shallow water connected with Lake Erie.

JUNCACEÆ.

JUNCUS, L. Rush. Bog-Rush.

J. ACUMINATUS, Michx.
Florence ; rare.

J. ALPINUS INSIGNIS, Fries.
Castalia, Oxford, shores of Lake Erie; frequent.
Kelley's Island.

J. BALTICUS LITTORALIS, Engelm.
Castalia, Cedar Point, Marblehead sand spit;
locally abundant.

J. BUFONIUS, L.
Sandusky near B. & O. and L. S. & M. S. Ry's;
rare.

J. CANADENSIS.
Shinrock and Sandusky where the so-called variety *longicaudatus* grows.

J, CANADENSIS BRACHYCEPHALUS, Engelm.
Castalia, Willow Point, Sandy Beach.

J. EFFUSUS, L. Common or Soft Rush.
Frequent. North Bass.

J. MARGINATUS, Rostk.
Berlin, Vermillion, east of Milan ; infrequent.

J. NODOSUS, L.
Frequent.

J. NODOSUS MEGACEPHALUS, Torr.
  Frequent. Islands.
J. SCIRPOIDES, Lam.*
  Oxford, southern Perkins, Vermillion; infrequent.
J. TENVIS, Willd.
  Common.

LUZULA, D C. Wood-Rush.

L. CAMPESTRIS, D C.
  Frequent, especially in Milan.
L. VERNALIS, D C.
  Vermillion River, Chapelle Creek; scarce.

LILIACEÆ.

ALETRIS, L.

A. FARINOSA, L.*
  Perrin's, Milan and Joseph Smith's, Perkins; rare

ALLIUM, L.

A. CANADENSE, L. Wild Garlic.
  Infrequent. Kelley's Island.
A. CERNUUM, Roth. Wild Onion.
  Common on the Islands, Peninsula, and at
  Castalia.
A. TRICOCCUM, Ait. Wild Leek.
  Islands, Peninsula, Florence; infrequent.

ASPARAGUS, L.

A. officinalis, L. Garden Asparagus.
  Escaped in many places. Islands.

CAMASSIA, Lindl.

C. FRASERI, Torr. Wild Hyacinth.
  Infrequent, but occurs on eight islands and in
  eight townships.

CHAMAELIRIUM. Willd.

C. CAROLINIANUM, Willd. Blazing-Star.
Southern Perkins, Margaretta Ridge, east of Milan, Berlin Heights ; rare.

DISPORUM, Salisb.

D. LANUGINOSUM, Nichols.
Flórence, Berlin ; scarce.

ERYTHRONIUM, L.

E. ALBIDUM, Nutt. White Dog's-tooth Violet.
A weed in vineyards west of Sandusky.
Common on Huron River bottoms, Infrequent or rare in other parts of the county.
Johnson's Island, Kelley's, Rattlesnake, Port Clinton.

E. AMERICANUM, Ker. Yellow Adder's-tongue.
Common.

HEMEROCALLIS, L.

H. *fulva*, L.
Roadsides; infrequent. North Bass.

LILIUM, L.

L. CANADENSE, L. Wild Yellow Lily.
Infrequent. Kelley's, Island.

L. PHILADELPHICUM, L. Wild Orangered Lily. Wood Lily.
Scarce.

L. SUPERBUM, L. Turk's-cap Lily.
Milan, Florence, Vermillion; rare. Mr. Haise of Florence found "several years ago a lily with forty or fifty flowers."

MAIANTHEMUM, Wigg.

M. CONVALLARIA, Wigg. (M. CANADENSE, Desf.) False Lily-of-the-valley.
Cedar Point and high banks of Old Woman Creek, Chapelle Creek and Vermillion River; infrequent.

MEDEOLA, L. Indian Cucumber-root.

M. VIRGINICA, L.

Florence, Berlin, Milan, Perkins; scarce.

OAKESIA, Watson.

O. SESSILIFOLIA, Watson.

Florence; rare.

ORNITHOGALUM, L. Star-of-Bethlehem.

O. *umbellatum* L.

Perkins, Sandusky, Put-in-Bay; rare.

POLYGONATUM, Adans.

P. BIFLORUM, Ell.   Smaller Solomon's Seal.
Common.

P. GIGANTEUM, Dietrich.   Great Solomon's Seal.
Common.

SMILACINA, Desf. False Solomon's Seal.

S. RACEMOSA, Desf.   False Spikenard.
Common.

S. STELLATA, Desf.

Common; less so on the mainland than the
preceding.

SMILAX, L. Greenbrier.

S. ECIRRHATA, Watson.

Perkins, Groton, Catawba, Kelley's Island; scarce.

S. HERBACEA, L.   Carrion-Flower.
Common.

S. HISPIDA, Muhl.

Frequent.   Islands.

S. ROTUNDIFOLIA, L.   Horse-brier.

Infrequent.   Put-in-Bay.

The "variety" *crenulata* S. & H. found at
Chapelle Creek.

TRILLIUM, L. Wake Robin.

T. ERECTUM, L.
Common.

T. GRANDIFLORUM, Salisb.
Common.

T. SESSILE, L.
Vermillion River flats ; frequent.

UVULARIA, L. Bellwort.

U. GRANDIFLORA.
Infrequent. Islands.

ZYGADENUS, Michx.

Z. ELEGANS, Pursh.*
Marblehead ; rare.

AMARYLLIDACEÆ.

HYPOXIS, L. Star-Grass.

H. ERECTA, L.
Infrequent.

DIOSCOREACEÆ.

DIOSCOREA, L. Yam.

D. VILLOSA, L. Wild Yam-root.
Frequent. Kelley's Island, Put-in-Bay.

IRIDACEÆ.

IRIS, L. Flower-de-Luce.

[I. CRISTATA, Ait, Crested Dwarf Iris.
Our specimens of this rare plant were collected
along the Vermillion River in what was said to
be Erie County, but the spot proves to be a few
yards south of the line. Eli Beecher, who owns
the adjacent flats in Erie County, says he has
seen it there.]

I. VERSICOLOR, L.   Larger Blue Flag.
Frequent. Islands.

SISYRINCHIUM, L.   Blue-eyed Grass.

S. ANGUSTIFOLIUM, Mill.
Infrequent.

S. GRAMINOIDES, Bicknell.
Infrequent.

ORCHIDACEÆ.

APLECTRUM, Torr.   Putty-Root.   Adam-and-Eve.

A. HYEMALE, Torr.
Rare. Puckrin's woods, Perkins.
"Smith's, Perkins," Ross Ransom. "Cedar Point,"
Claassen and Krebs. "Marblehead," Gertrude
Johnson. "Vermillion," Otto Todd. "Formerly
considerable near the quarry on west branch of
Vermillion River," Eli Beecher.

CALOPOGON, R. Br.

C. PULCHELLUS, R. Br.
South-west of Castalia; rare. Seen only in 1895.

CORALLORHIZA, Haller.   Coral-root.

C. MULTIFLORA, Nutt.
Florence, Huron, Catawba; rare.

C. ODONTORHIZA, Nutt.
Blair Creek, Florence: Graham's woods, Huron;
Smith's woods, Perkins; rare.

CYPRIPEDIUM, L.   Lady's Slipper.   Moccason-flower.

C, CANDIDUM, Muhl.*   Small White Lady's Slipper.
Along a railroad near Castalia; locally common.

C. PUBESCENS. Willd.   Larger yellow Lady's Slipper.
In seven townships, but rare.

C. SPECTABILE, Salisb.   Showy Lady's Slipper.
One spot on high, wet, shale bank of east branch,
Vermillion River. An orchid found by Job Fish
"about 1859, the most beautiful wild flower" he
"ever found" was probably of this species.

GOODYERA, R. Br. Rattlesnake-Plantain.

G. PUBESCENS, R Br.
   Florence, Berlin, Milan, Oxford, Perkins; scarce.

HABENARIA, Willd. Rein-Orchis.

H. BRACTEATA, R. Br.
   In five townships; rare.

H. HERBIOLA, R. Br. (H. VIRESCENS, Spreng.)
   In five townships; rare.

H. HOOKERIANA, Torr.
   "Margaretta Ridge," Henry Schoepfle; one plant.

H. LACERA R. Br. Ragged Fringed-Orchis.
   Perkins, Milan, Vermillion; rare.

H. PSYCODES, Gray. Purple Fringed-Orchis.
   Florence, Milan, "Cedar Point," Leslie Stair: rare.

H. TRIDENTATA, Hook.
   East of Milan; one plant.

LIPARIS, Richard. Twayblade.

L. LŒSELII, Richard.
   Bog near Cedar Point Light House,

ORCHIS, L.

O. SPECTABILIS, L. Showy Orchis.
   Rather frequent in Florence, infrequent in four townships.

POGONIA, Juss.

P. PENDULA, Lindl.
   "Florence," Josephine Fish. also Otto Todd; East Berlin; "Perkins," Ransom; local.

SPIRANTHES, Richard. Ladies' Tresses.

S. CERNUA, Richard.
   Local. This and *Orchis spectabilis* are less rare than our other orchids.

S. GRACILIS, Beck.
   "Bloomingville," W. A. Kellerman. Perkins; rare.

DICOTYLEDONES.

SAURURACEÆ.

SAURURUS, L. Lizard's-tail.

S. CERNUUS, L.
Frequent in eastern part of the county; infrequent in Huron, Milan and Perkins.

JUGLANDACEÆ.

CARYA, Nutt. Hickory.

C. ALBA, Nutt. Shell-bark or Shag-bark Hickory.
Abundant. Hickory is used in Sandusky by two wheel works and two whip-stalk factories; also by the Sandusky Tool Company for chisel handles, for tin-smith's mallets, and for ladder-rounds that are sent to Northern Michigan for use in the copper mines.

C. AMARA, Nutt. Bitter-nut or Swamp Hickory.
Frequent. One in the German Settlement has a circumference of 9 feet, 8 inches.

C. MICROCARPA, Nutt.
Frequent, at least in Perkins.

C. SULCATA, Nutt. Big Shell-bark. King-nut.
Frequent.

C. TOMENTOSA,   Nutt.   Mocker-nut.   White-heart
Hickory.
Frequent.   Put-in-Bay.
C. PORCINA, Nutt.   Pig-nut or Broom Hickory.
Frequent.   Islands.

JUGLANS, L.

J. CINEREA, L.   Butternut.   White Walnut.
Infrequent.
J. NIGRA, L.   Black Walnut.
Frequent.   Said to have grown formerly on
Kelley's Island, and Middle Bass.   The number
and size of the walnut stumps along the border
of the Huron marsh east of Sandusky and of the
prostrate trunks in the marsh is remarkable.
See page 14.

SALICACEÆ.

POPULUS, L.

P. *alba*, L.   White Poplar.   Abele.
Frequent in the vicinity of planted trees.
Kelley's Island.   Put-in-Bay.
P. GRANDIDENTATA, Michx.   Large-toothed Aspen.
Rather frequent.   Put-in-Bay.   Plentiful along
the lake shore drive east of Huron.
P. HETEROPHYLLA, L.   Downy Poplar.
Florence, Huron ; rare.
P. MONILIFERA, Ait.   Cotton-wood.   Necklace Poplar.
Common.
P. TREMULOIDES, Michx.   American Aspen.
Frequent, especially on the Islands.

SALIX, L.   Willow.   Osier.

S. *alba cærulea*, Koch.   Blue Willow.
Cedar Point and Sandusky near the Bay ; rare.

S. *alba vitellina*, Koch.   Golden Osier.
    Frequent.   Islands.

S. AMYGDALOIDES, Anders.
    Frequent.

S. CANDIDA, Willd.*   Sage Willow.   Hoary Willow.
    Castalia prairie ; rare.

S. CORDATA, Muhl.   Heart-leaved Willow.
    Common, but not noticed on Kelley's Island.

S. CORDATA ANGUSTATA, Anders.
    Infrequent.   Put-in-Bay.

S. DISCOLOR, Muhl.   Glaucous Willow.
    Frequent, as is also the " variety " *eriocephala*.

S. GLAUCOPHYLLA, Bebb.
    Cedar Point, Castalia ; infrequent.

S. HUMILIS, Marsh.   Prairie Willow.
    Oxford ; scarce.

S. LONGIFOLIA, Muhl.
    Common, especially along the lake.

S· LUCIDA, Muhl.
    Florence, Marblehead, Put-in-Bay ; infrequent.

S. NIGRA, Marsh.   Black Willow.
    Frequent.   Islands.

S. NIGRA FALCATA, Torr.
    Frequent.

S. PETIOLARIS, Smith.
    House's swamp, southern Perkins.

S. *purpurea*, L.   Purple Willow.
    Infrequent.   Kelley's Island.   Put-in-Bay.

S. ROSTRATA, Richardson.
    Infrequent.   Islands.

S. SERICEA, Marsh.   Silky Willow.
    House's swamp, Perkins.   Milan ?

S. *fragilis* × *alba*.
    Castalia, etc.

BETULACEÆ.

CARPINUS, L. Iron-wood.

C, AMERICANA, Michx. American Hornbeam. Blue or
Water Beech.
Frequent. "Formerly many on Kelley's Island."
Lester Carpenter.

CORYLUS, L.

C. AMERICANA, Walt. Hazel-nut.
Common. Not on the Islands.

OSTRYA, L. Iron-wood.

O. VIRGINICA, Willd. American Hop-Hornbeam. Lever-
wood.
Common, especially on rocky shores of the
Islands.

FAGACEÆ.

CASTANEA, L.

C. SATIVA AMERICANA, Watson. Chestnut.
Common in Erie County in sandy soil.
Chestnut fence posts sometimes put forth leafy
shoots.

FAGUS, L.

F. FERRUGINEA, Ait. American Beech.
Not on Islands or Peninsula, nor within five miles
of Sandusky. A few in Kromer's woods and
farther south along Pipe Creek. Infrequent along
Huron River in Milan, frequent in Berlin, common in Vermillion, abundant in Florence. "Two
trees on Put-in-Bay thirty years ago," Vroman.
"Formerly a few on Middle Bass." Wood
found in the submerged forest, Huron Marsh.
Most Sandusky children do not know beech
nuts. Wood used by Sandusky Tool Company
for planes.

QUERCUS, L. Oak.

Q. ALBA, L. White Oak.

Common.

Q. BICOLOR, Willd. Willd. Swamp White Oak.

Frequent. Kelley's Island.

Q. COCCINEA, Wang. Scarlet Oak.

East of Milan; frequent. Marblehead, Port Clinton, Catawba and probably elsewhere.

Q. IMBRICARIA, Michx. Laurel or Shingle Oak.

Common in middle and western parts of Erie County. Abundant in Oxford and on Cedar Point.

Q. MACROCARPA, Michx. Bur Oak, Over-cup or Mossy-cup Oak.

Frequent. Islands. Under the large Bur Oak at the corner of Wayne and Jefferson Sts., the Indians used to hold their councils. It is said to have grown very little since the early settlers came to Sandusky.

Q. MUHLENBERGII, Engelm. Yellow Oak. Chestnut Oak.

Common on the Peninsula and Islands. Less frequent in Erie County.

Q. PALUSTRIS, Du Roi. Swamp Spanish or Pin Oak.

Common. Not noticed on the Islands.

Q. PRINUS, L. Rock Chestnut Oak.

Sandusky. Marblehead, Islands and elsewhere? The oak in Judge Mackey's yard on Columbus Ave. south of the fair grounds is of this species.

Q. RUBRA, L. Red Oak.

Common.

Q. VELUTINA, Lam. (Q. TINCTORIA,) Bartram. Black Oak. Quercitron.

Common. Kelley's and Put-in-Bay the only islands. On Cedar Point, where this species abounds, is a tree which I should call Q. *marylandica*, Muench., were I not advised differently, and other trees of the same sort or else hybrids between it and Q. *velutina*. None of these were noticed until September, 1898.

ULMACEÆ.

CELTIS, L.

C. OCCIDENTALIS, L. Hackberry. Sugar-berry.
Frequent. Common on the Islands and Cedar
Point.

ULMUS, L.

U. AMERICANA, L. American or White Elm.
Common. Wood used for the handles and bands
of baskets and for lime barrels.

U. FULVA, Michx. Slippery or Red Elm.
Frequent. All the Islands.

MORACEÆ.

CANNABIS, L.

C. sativa, L. Hemp.
Roadside, Margaretta or Groton ; very rare.

HUMULUS, L.

H. LUPULUS, L. Hop.
Castalia, Milan ; infrequent.

MACLURA, Nutt.

M. AURANTIACA, Nutt. Osage Orange.
Found only near where it has been planted ;
scarcely naturalized. The row of trees on the
Ransom place, Castalia road, probably surpasses
any farther north in America.

MORUS, L.

M. alba, L.* White Mulberry.
Rare in woods, where the seeds have probably
been dropped by birds.

M. RUBRA, L. Red Mulberry.
Throughout, but infrequent. "Formerly common
at Port Clinton." Islands.

URTICACEÆ.

BŒHMERIA, Jacq.

B. CYLINDRICA, Sw. False Nettle.
Common.

LAPORTEA, Gaudichaud.

L. CANADENSIS, Gaudichaud. Wood-Nettle.
Common.

PARIETARIA, L.

P. PENNSYLVANICA, Muhl. Pellitory.
Abundant.

PILEA, Lindl.

P. PUMILA, Gray. Richweed. Clearweed.
Common. Kelley's the only island.

URTICA, L. Nettle.

U. GRACILIS, Ait.
Common.

SANTALACEÆ.

COMANDRA, Nutt. Bastard Toad-flax.

C. UMBELLATA, Nutt.
Frequent.

ARISTOLOCHIACEÆ.

ARIRTOLOCHIA, L.

A. SERPENTARIA, L. Virginia Snakeroot.
Florence, Berlin, Perkins, Margaretta; scarce.

ASARUM, L. Wild Ginger.

A. ACUMINATUM, Bicknell.
Florence and probably elsewhere.

A. REFLEXUM, Bicknell.
Huron River, Milan, and probably elsewhere.
The variety *ambiguum* also occurs.

POLYGONACEÆ.

FAGOPYRUM, Gaertn.

F. ESCULENTUM, Moench. Buckwheat.
Infrequent, except in fields where it has sometime
been sown.

POLYGONUM, L. Knotweed.

P. ACRE, H. B. K. Water Smartweed.
Common.

P. AMPHIBIUM, L.*
Marblehead ; rare.

P. ARIFOLIUM, L. Halberd-leaved Tear-thumb.
Bristol's woods, Florence.

P. AVICULARE, L. Knot-grass. Door-weed.
Abundant.

P. CAREYI, Olney.*
Southern Perkins.

P. *convolvulus*, L. Black Bindweed.
Common.

P. DUMETORUM, L. Copse or Hedge Buckwheat.
Milan, Marblehead. This or P. SCANDENS is
common and grows on the Islands.

P ERECTUM, L. Erect Knotweed.
Common.

P. HARTWRIGHTII Gray.*
A few plants near L. S. & M. S. freight house.
Doubtless introduced.

P. HYDROPIPER. Smart-weed. Water Pepper.
Common.

P HYDROPIPEROIDES, Michx. Mild Water Pepper.
Infrequent. Kelley's Island.

P. INCARNATUM Ell.
Frequent in wet places near Lake Erie and Sandusky Bay, also at Castalia.

P. LAPATHIFOLIUM, L.
Cedar Point, Lockwood's ; infrequent.

P. LITTORALE, Link.*
Sandusky; frequent. Kelley's Island, and probably many other places near Lake Erie. We failed to distinguish it, till recently, from P. *aviculare*.

P. MUHLENBERGII, Watson.
Frequent. Islands.

P, *orientale*, L.
Barely naturalized in two or three places.

P. PENNSYLVANICUM, L.
Abundant. Kelley's and Middle Bass the only islands where it has been noticed.

P. *persicaria*, L. Lady's Thumb.
Abundant.

P. RAMOSISSIMUM, Michx.*
Hill's woods, southern Perkins ; one plant.

P. SAGITTATUM, L. Arrow-leaved Tear-thumb.
Frequent.

P. SCANDENS, L. Climbing False Buckwheat.
Margaretta, Cedar Point and probably elsewhere. See P. *dumetorum*.

'P. TENUE, Michx.*
Marblehead ; frequent. Margaretta, between quarry and Castalia road. Only in thin soil overlying the lime stone.

P. VIRGINIANUM, L.
Common. Not on the Islands.

RUMEX, L.

R. *acetosella*, L. Field or Sheep Sorrel.
Abundant. Put-in-Bay ; rare. "Kelley's Island." Not on other islands.

R. ALTISSIMUS, Wood.   Pale Dock.
> Sandusky by Big Four track, Put-in-Bay ; rare ;
> also Oak Harbor, Ottawa County.

R. BRITANNICA, L.   Great Water-Dock.
> Marshes connected with Sandusky Bay; frequent.

R. *crispus*, L.   Curled Dock.
> Abundant.

R. *obtusifolius*, L.   Bitter Dock.
> Common.

R. VERTICILLATUS.
> Common in marshes.

CHENOPODIACEÆ.

ATRIPLEX, L.   Orache.

A. ARGENTEA, Nutt.
> Near Big Four R. R., Sandusky and Castalia ;
> rare.

A. HASTATA, L.
> Common near Lake and Bay.   In many places in
> Sandusky the most common weed.

A. LITTORALIS, L.*
> Sandusky ; frequent.   Huron.

CHENOPODIUM, L.   Pigweed.

C. *album*, L.   Lamb's Quarters.   Pigweed.
> Common.

C. *album viride*, Moq.
> Common.

C. *ambrosioides*, L.   Mexican Tea.
> L. S. & M. S. R. R. yards, Sandusky ; rare.

C. BOSCIANUM, Moq.
> Cedar Point, Perkins, Kelley's Island, and,
> doubtless, elsewhere.

C. *botrys*, L. Jerusalem Oak. Feather Geranium.
Western part of Erie Co., mostly along railways
(C. S. & H. and L. E. & W). Marblehead.
Kelley's Island. Infrequent except on Marble-
head.

C. *glaucum*, L. Oak-leaved Goosefoot.
Castalia prairie and along L. E. & W. Ry. at
Castalia and Sandusky; rare.

C. HYBRIDUM, L. Maple-leaved Goosefoot.
Islands, Peninsula, Cedar Point, Perkins,
Margaretta; frequent.

C. LEPTOPHYLLUM, Nutt.*
Cedar Point and probably elsewhere; infrequent.

C. *murale*, L.
Sandusky; infrequent.

C. *urbicum*, L.
Rather frequent on the Peninsula, and in the
western third of Erie Co. Kelley's Island.

AMARANTACEÆ.

ACNIDA, L.

A- TUBERCULATA, Moq.
Wet ground near Lake and Bay and at Castalia;
infrequent. Kelley's Island. Middle Bass

AMARANTUS, L. Amaranth.

A. ALBUS. L. Tumble Weed.
Common.

A. BLITOIDES, Watson.
Common.

A. *chlorostachys*, Willd.
Common.

A. *hypochondriacus*, L.
Sandusky, Perkins; scarce.

A. *paniculatus*, L.
Roadsides, Sandusky and Islands; infrequent.
A. *retroflexus*, L.
Common.

PHYTOLACCACEÆ.

PHYTOLACCA, L.

P. DECANDRA, L. Poke. Scoke. Pigeon-berry. Garget. Common.

NYCTAGINACEÆ.

OXYBAPHUS, Vahl.

O. NYCTAGINEUS, Sweet.
L. S. & M. S. Ry. in eastern Sandusky.

AIZOACEÆ.

MOLLUGO, L.

M. VERTICILLATA, L. Carpet-weed.
Sandusky, southern Perkins, Milan; local.

PORTULACACEÆ.

CLAYTONIA, L.

C. VIRGINICA, L. Spring Beauty.
Abundant.

PORTULACA, L.

P. *oleracea*, L. Purslane.
Abundant.

CARYOPHYLLACEÆ.

ANYCHIA, Michx. Forked Chickweed.

A. CAPILLACEA, DC.
Infrequent. Put-in-Bay.

A. DICHOTOMA, Michx.
Marblehead, Catawba, infrequent. Plentiful in places on the shale in Oxford and Perkins.

ARENARIA, L. Sandwort.

A. LATERIFLORA, L.
Lake woods, Port Clinton and Big woods, Perkins; rare.

A. *serpyllifolia*, L. Thyme-leaved Sandwort.
Islands, Peninsula, Margaretta, western Perkins; frequent.

A. STRICTA, Michx.
Islands, Peninsula, Margaretta, western Perkins, Cedar Point; locally common.

CERASTIUM, L. Mouse-ear Chickweed.

C. NUTANS, Raf.
Frequent. Islands.

C. OBLONGIFOLIUM, Torrey.*
More frequent than the last on Islands and Peninsula and in the western half of Erie Co.

C. *vulgatum*, L.
Common.

LYCHNIS, L.

L. *dioica*, L. Red Lychnis.
Avery; probably adventive.

L. *githago*, Scop. Corn Cockle.
Common. Kelley's the only Island.

L. *vespertina*, Sibth.
Franz Otto's, Perkins.

SAPONARIA, L.

S. *officinalis*, L.   Soapwort.   Bouncing Bet.
Frequent.   Islands.

SILENE, L.

S. ANTIRRHINA, L.   Sleepy Catchfly.
Frequent.   Kelley's Island.

S. *conica*, L.*   Corn Catchfly.
"Sandy field west of B. & O. R. R., southern
Perkins." Ross Ransom. The first recorded
appearance of this plant in the United States was
at Clyde, Sandusky County, where it was introduced in Crimson Clover seed, 1896.

S. *cucubalus*, Wibel.*   Bladder Campion.
Well established and increasing in a field of
James Hamilton, Kelley's Island.

S. *dichotoma*, Ehrh.   Forked Catchfly.
Northeast of Port Clinton; probably adventive.

S. *noctiflora*, L.   Night-flowering Catchfly.
Sandusky; scarce.

S. VIRGINICA, L.   Fire Pink.
Put-in-Bay; frequent. Kelley's Island. Ca-
tawba, Hartshorn's, Johnson's Island. "Cedar
Point," Alden Knight.

STELLARIA, L.

S. LONGIFOLIA, Muhl.   Long-leaved Stitchwort.
Frequent.

S. *media*, Cyrill.   Common Chickweed.
Abundant.

NYMPHÆACEÆ.

BRASENIA, Schreber.

B. PELTATA, Pursh.   Water-shield.
Cedar Point; one plant.

NELUMBIUM, Adans. Sacred Bean.

N. LUTEUM, Willd. American Nelumbo or Lotus. Water Chinkapin or Wankapin.

In still, deep, water at several places about Sandusky Bay, in the East and West Harbors, at Port Clinton where a large amount of it grows in the Portage River, and west to Monroe, Michigan, but believed to grow nowhere along the American shore of Lake Erie east of the mouth of the Old Woman Creek. A hundred acres of it at the head of Sandusky Bay and along the river, more, probably, than the whole quantity in the United States farther east. The lotus has the largest flowers and largest leaves of any plant in the Sandusky flora. Petioles sometimes 9 feet long; "blades 26 inches broad."

NUPHAR, Smith. Spatter-Dock.

N. ADVENA, Ait. Yellow Pond-Lily.

Sandusky Bay, Middle Bass, Blair Creek; freqnent.

NYMPHAEA, Tourn. Water-Lily.

N. TUBEROSA, Paine.

Common in still waters connected with the Bay and Lake.

CERATOPHYLLACEÆ.

CERATOPHYLLUM, L. Hornwort.

C. DEMERSUM, L.

Sandusky Bay, East Harbor, Port Clinton, Put-in-Bay; common.

MAGNOLIACEÆ.

LIRIODENDRON, L. Tulip-tree.

L. TULIPIFERA, L.

Scarce in the western but frequent in the eastern part of the county, where many of the largest trees in the primeval forest were of this species. Lakeside. Commonly called White-wood and improperly, Yellow Poplar and White Poplar. The wood suitable for pumps, troughs and hollow ware.

MAGNOLIA, L.

M. ACUMINATA, L. Cucumber-tree.

Two trees near the iron bridge across east branch of Vermillion River. "Big woods, Perkins."

AMONACEÆ.

ASIMINA, Adans.

A. TRILOBA, Dunal. North American Papaw.

Not found near Sandusky, but near Milan and in many places east from there to the Vermillion River, especially along the Old Woman Creek and other streams. Also in the forest west of Castalia in Sandusky Co. "Formerly on Kelley's Island."

RANUNCULACEÆ.

ACTÆA, L.

A. ALBA, Mill. White Baneberry.

Frequent.

A. SPICATA RUBRA, Ait. Red Baneberry.

Cedar Point, Perkins, Margaretta Ridge; scarce. "Berlin."

ANEMONE, L.

A. ACUTILOBA, Laws. (Hepatica acutiloba, D C.)
Liver-leaf.
Frequent. Islands.

A. CYLINDRICA, Gray. Long-fruited Anemone.
Infrequent but observed in eight townships.

A. DICHOTOMA, L. (A pennsylvanica, L.)
Common. All islands, except Kelley's.

A. HEPATICA, L. (Hepatica triloba, Chaix) Liver-leaf.
Frequent. Not observed in Florence where A.
acutiloba is rather common. Islands. Both
species more frequent on the Peninsula than in
Erie Co.

A. NEMOROSA, L. Wind-flower. Wood Anemone.
Common.

A. THALICTROIDES, L. Rue-Anemone.
Common. Sometimes double. In blossom as
late as September.

A. VIRGINIANA, L.
Frequent. Islands.

AQUILEGIA, L. Columbine.

A. CANADENSIS, L.
Not noticed near Sandusky, except on Cedar
Point, but common among rocks on the Penin-
sula and Islands and at Margaretta Ridge.
Berlin, Vermillion, Florence. Adorns the rocky
shores of the islands.

CALTHA, L. Marsh Marigold.

C. PALUSTRIS, L.
Frequent.

CIMICIFUGA, L. Bugbane.

C. RACEMOSA, Nutt. Black Snakeroot. Black Cohosh.
Common in woods in eastern part of Erie Co,
and extending west to Perkins.

CLEMATIS, L. Virgin's Bower

C. VIRGINIANA, L.
> Frequent. North Bass.

DELPHINIUM, L. Larkspur.

D. *ajacis*, L.
> Spontaneous in gardens and near them.

D. AZUREUM, Michx.
> One plant found by L. S. & M. S. Ry. between Venice and Bay Bridge, by Will Newberry. Probably adventive.

HYDRASTIS, Ellis. Orange-root.

H. CANADENSIS, L. Golden Seal.
> Frequent in rich woods long undisturbed. "Kelley's Island." "Catawba."

ISOPYRUM, L.

I. BITERNATUM, Torr & Gray.
> Vermillion River, southeren Florence; scarce. "Huron River at Norwalk" Leslie D. Stair.

NIGELLA, L.

N. *damascena*, L. Fennel-flower.
> Spontaneous in gardens and rarely escaped.

RANUNCULUS, L. Crowfoot. Buttercup.

R. ABORTIVUS, L. Small-flowered Crowfoot.
> Common.

R. *acris*, L. Tall or Meadow Buttercup.
> Florence, Berlin, Huron, Sandusky, Put-in-Bay; infrequent.

R. CIRCINATUS, Sibth. Stiff Water Crowfoot.
> Sandusky Bay, Castalia, Mill's Creek; frequent.

R. FASCICULARIS, Muhl. Early Buttercup.
> Margaretta, Huron, Peninsula, Johnson's Island, Kelley's Island; locally plentiful.

R. MULTIFIDUS, Pursh.

House's swamp, Perkins; Castalia; Peninsula; Islands; infrequent.

R. OBTUSIUSCULUS, Raf. (R. ambigens, Watson,) Water Plantain Spearwort.

Millan and Florence; rare.

R. PENNSYLVANICUS, L. f. Bristly Buttercup.

Sandnsky and Willow Point near the Bay, Catawba; rare.

R. RECURTATUS, Poir. Hooked Crowfoot.

Frequent, especially along rivers.

R. SCELERATUS, L. Cursed Crowfoot.

Frequent. Islands.

R. SEPTENTRIONALIS, Poir. Swamp or Marsh Buttercup.

Common. Kelley's and "Put-in-Bay" the only islands.

THALICTRUM, L. Meadow-Rue.

T. DIOICUM, L. Early Meadow-Rue.

Common.

T. POLYGAMUM, Muhl. Tall Meadow-Rue.

Frequent.

T. PURPURASCENS, L. Purplish Meadcw-Rue.

Frequent, especially near Castslia.

BERBERIDACEÆ.

BERBERIS, L. Barberry.

B. VULGARIS, L. Common Barberry.

Woods, Milan and Huron; rare. Seeds probably dropped by birds.

CAULOPHYLLUM, Michx. Blue Cohosh.

C. THALICTROIDES, Michx.

Florence, Vermillion, Berlin, Perkins, Johnson's Island; infrequent.

JEFFERSONIA, Barton. Twin-leaf.

J. BINATA, Barton, (J. DIPHYLLA, Pers.)
Johnson's Island, but nowhere else near
Sandusky. Lockwood's woods, Peninsula.
Several places along Vermillion River, Florence.

PODOPHYLLUM, L. Mandrake.

P. PELTATUM, L. May-Apple.
Abundant. Fruit edible. "Leaves and roots
poisonous." Gray.

MENISPERMACEÆ.

MENISPERMUM, L, Moonseed.

M. CANADENSE, L.
Frequent. Islands.

LAURACEÆ.

LINDERA, Thumb.

L. BENZOIN, Meisn. Spice-bush. Benjamin-bush.
In rich woods in Erie County the most abundant
shrub.

SASSAFRAS, Nees.

S. OFFICINALE, Nees.
Frequent. "Formerly on the Islands." Submerged trunks found in Huron Marsh. See page
15. Some trees on the Peninsula measured by
J. R. Kelly have trunks with circumferences as
follows: 8 ft. 1 in.; 7½ft.; 6 ft. 10 in.; 6 ft.
Formerly sassafras oil was made in Sandusky.

PAPAVERACEÆ.

CHELIDONIUM, L. Celandine.

C. *majus*, L.
Scarce.

PAPAVER, L. Poppy.

P. *argemone*, L. Rough-fruited Corn-Poppy.
"In a Crimson Clover field, Perkins." Ross
Ransom. Probably adventive.

P. *somniferum*, L. Opium Poppy.
Along a railroad, Sandusky; rare and
adventive.

SANGUINARIA, Dell. Blood-root.

S. CANADENSIS, L.
Frequent. Islands.

FUMARIACEÆ.

CORYDALIS, Vent.

C. AUREA, Willd. Golden Corydalis.
"Port Clinton," Leslie D. Stair.

C. FLAVULA, D C.
Peninsula and Islands including Johnson's.
"Cedar Point." Krebs.

DICENTRA, Borkh.

D. CANADENSIS, Walp. Squirrel Corn.
Berlin, Florence, Milan, Perkins; rare.
"Vermillion" Otto K. Todd.

D. CUCULLARIA, Bernh. Dutchman's Breeches.
Frequent. All the Islands.

FUMARIA, L. Fumitory.

F. *officinalis, L.*
Sandusky, Cedar Point, Kelley's Island; rare.

CRUCIFERÆ.

ALYSSUM, L.

A. CALYCINUM, L.
"Catawba" Nettie Schnaitter.

ARABIS, L. Rock Cress.

A. CANADENSIS, L. Sickle-pod.
Perkins, Margaretta, Peninsula, Johnson's Island, Put-in-Bay. Middle Bass; infrequent.

A. DENTATA, Torr & Gray.
Cedar Point, Florence, Johnson's Island, North Bass, Green Island; infrequent.

A. DRUMMONDII, Gray. (A. CONFINIS. Watson.)
Cedar Point and Islands; frequent.

A. HIRSUTA, Scop.
Marblehead; common. Catawba. Mouse Island, Margaretta, Huron River.

A. LÆVIGATA. DC.
Frequent. Islands.

A. LYRATA. L.
Cedar Point; common. Perkins, Marblehead.

A. PERFOLIATA, Lam. Tower Mustard.
Johnson's Island; rare.

BARBAREA, R. Br. Winter Cress.

B. VULGARIS, R. Br. Yellow Rocket.
Frequent, Green Island. Some of the specimens, at least, belong to the "variety" *stricta*, which may be distinct.

BRASSICA, L.

B. *napus*, L. Rape.
Sandusky, Vermillion; adventive.

B. *nigra*, Kock. Black Mustard.
Common.

B. *sinapistrum*, Boiss. Charlock.
Abundant.

CAKILE, Tourn. Sea-Rocket.

C. MARITIMA, Scop. (C. AMERICANA, Nutt.)
Shores of Lake and Bay; common.

CAMELINA, Crantz. False Flax.

C. *sativa*, Crantz.
Sandusky and Avery; rare.

CAPSELLA, Medic. Shepherd's Purse.

C. *bursa-pastoris*, Medic.
Abundant.

CARDAMINE, L. Bitter Cress.

C. DIPHYLLA, Wood. Two-leaved Toothwort.
Huron River near Millan; rare. Florence; scarce. "Berlin Heights" Chas. Judson.

C. LACINIATA, Wood. Toothwort. Pepperroot.
Common.

C. PENNSYLVANICA, Muhl.
Frequent. Kelley's Island. North Bass.

C. RHOMBOIDEA, DC. Spring Cress.
. Common.

C. RHOMBOIDEA PURPUREA, Torr.
Common.

COCHLEARIA, L.

C. *armoracia*, L. (Nasturtium *arnoracia*, *Fries*.)
Horseradish.
Frequent. Islands.

CONRINGIA, Link.

C. *orientalis*, Dum.* Hare's-ear Mustard.
Four plants found along railroad near ice houses, eastern Sandusky, 1897, by Geo. Gilbert.

DRABA, Dill. Whitlow-Grass.

D. CAROLINIANA, Walt.
Common on Marblehead and in some places in Margaretta in thin soil overlying the limestome.

D. *verna*, L.
"Perkins," Lindsey House. rare.

ERYSIMUM, L. Treacle Mustard.

E. PARVIFLORUM, Nutt.*
> One place along L. E. & W. Ry., west of Castalia; rare.

LEPIDIUM, L Pepperwort. Peppergrass.

L. APETALUM, Willd. (L. INTERMEDIUM, Gray.)
> Sandusky; infrequent.

L. *campestre*, R. Br.
> Sandusky, Perkins, Margaretta, Peninsula, Kelley's Island, Put-in-Bay. Common in places, especially on the Peninsula.

L. VIRGINICUM, L. Wild Peppergrass.
> Common.

NASTURTIUM, R. Br. Water-Cress.

N. LACUSTRE, Gray. Lake Cress.
> Shinrock; rare.

N. *officinale*, R. Br. True Water-Cress.
> Castalia; frequent.

N. PALUSTRE, D C. Marsh Cress.
> Common. On the Islands, and generally near the Lake or Bay, the variety *hispidum* is more common.

N. *sylvestre*, R. Br.* Yellow Cress.
> Four places in Perkins, three of them near or not far from Pipe Creek.

SISYMBRIUM, L.

S. *alliaria*, Scop.
> "Kelley's Island." Probably adventive.

S. CANESCENS, Nutt. Tansy Mustard.
> Cedar Point, Marblehead, Islands; frequent.

S. *officinale*, Scop. Hedge Mustard.
> Common.

THLASPI, L.

T. *arvense*, L. Field Pennycress.
> Sandusky; rare and adventive.

CAPPARIDACEÆ.

CLEOME, L.

C. GRAVEOLENS, Raf. (POLANISIA GRAVEOLENS, Raf.)
Common on sandy beaches. Also in gravel along
L. E. & W. R. R.

RESEDACEÆ.

RESEDA, L. Mignonette.

R. *lutea*, L.
Sandusky, Kelley's Island; rare and adventive.

DROSERACEÆ.

DROSERA, L. Sundew.

D. ROTUNDIFOLIA, L.
East of Milan; very rare.

CRASSULACEÆ.

PENTHORUM, Gronov. Ditch Stone-crop.

P. SEDOIDES, L.
Frequent. Islands.

SEDUM, L. Stone-crop. Orpine.

S. *acre*, L. Mossy Stone-crop.
Kelley's Island, roadside by the cemetery. Cedar
Point near the Light House. Escaped.
S. *telephium*, L. Orpine. Live-for-ever.
Bogart, Castalia, and Sandhill cemeteries. Put-
in-Bay, North Bass, "Marblehead" U G. Sanger
S. TERNATUM, Michx. Wild Stone-crop.
Frequent at the foot of steep shale banks of
streams. Put-in-Bay. Gibraltar.

SAXIFRAGACÆ.

CHRYSOSPLENIUM, L. Golden Saxifrage.

C. AMERICANUM, Schwein.
Vermillion River, Florence; two places.

HEUCHERA, L. Alum-root.

H. AMERICANA, L.
Common.

MITELLA, L. Bishop's-Cap. Mitrewort.

M. DIPHYLLA, L.
Infrequent.

PARNASSIA, L. Grass of Parnassus.

P. CAROLINIANA, Michx.
Castalia; frequent. Perkins, Milan, Florence;
rare.

PHILADELPHUS, L.

P. *coronarius*, L. Mock Orange. Syringa.
Sparingly escaped at Sandusky and Berlin
Heights.

SAXIFRAGA, L. Saxifrage.

S. PENNSYLVANICA, L. Swamp Saxifrage.
Milan and Florence ; scarce.

TIARELLA, L. False Mitrewort.

T. CORDIFOLIA, L.
East fork, Vermillion River; rare.

GROSSULARIACEÆ.

RIBES, L.

R. AUREUM, Pursh. Missouri or Buffalo Currant.
Well established on south side of Kelley's Island.
Roadside near a house in Margaretta.

R. CYNOSBATI. L, Gooseberry.
Common.

R. FLORIDUM, L'Her. Wild Black Currant.
   Infrequent. Kelley's Island.
R. LACUSTRE, Poir.
   "Cedar Point." Millie Carter.

HAMAMELIDACEÆ.

Hamamelis, L. Witch-Hazel.

H. VIRGINIANA, L.

Florence, Vermillion, Berlin, Milan; frequent.
"Portage River."

PLATANACEÆ.

PLATANUS, L. Sycamore.

P. OCCIDENTALIS, L. Buttonwood.
   Frequent. Islands. The largest tree in Erie
   county is probably the buttonwood six miles
   south of Sondusky, in the woods, but near the
   road and a little east of Pipe Creek.

ROSACEÆ.

AGRIMONIA, L. Agrimony.

A. EUPATORIA, L.
   Common. Kelley's the only Island.
A. MOLLIS, Torr. & Gray.
   Perkins and doubtless elsewhere.
A. PARVIFLORA, Soland.
   Frequent. In places, abundant.
A. STRIATA Michx.
   Margaretta Ridge. Probably elsewhere.

FRAGARIA, L. Strawberry.

F. VESCA, L.
   Peninsula, Kelley's Island, Put-in-Bay, Cedar
   Point, Margaretta, Berlin; frequent in rocky
   places.

F. VIRGINIANA, Duchesne.
Common. Kelley's, Put-in-Bay and Mouse the only Islands. Many specimens answer to description of the "variety" *illinoense.*

GEUM, L. Avens.

G. ALBUM, Gmelin.
Common.

G. STRICTUM, Soland.
Southern Perkins; rare.

G; VERNUM, Torr. & Gray.
Johnson's Island, Marblehead, Berlin, Perkins, etc.; rather frequent.

G. VIRGINIANUM, L.
Frequent. Kelley's Island? Put-in-Bay.

NEILLIA, D. Don. Ninebark.

N. OPULIFOLIA, Benth. & Hook.
Common on rocky shores of Peninsula and Islands. Vermillion River; rare.

POTENTILLA, L. Cinquefoil.

P. ANSERINA, L. Silver-weed.
Common on sandy shores of Lake and Bay, back a few yards from the water. Middle Bass, North Bass, Rattlesnake Island.

P. ARGUTA, Pursh.
Marblehead, Port Clinton, Put-in-Bay, Margaretta Ridge, Krieger's, Perkins; infrequent.

P. CANADENSIS, L. Five-finger.
Common. Not on the Islands.

P. FRUTICOSA, L. Shrubby Cinquefoil.
Castalia prairie; common. In blossom as late as October 10th.

P. NORVEGICA, L.
Frequent. In places abundant. Put-in-Bay.

P. SUPINA, L.
Huron and several places about Sandusky Bay.

ROSA, L. Rose.

R. BLANDA, Ait.

Cedar Point, Oxford, Groton, Margaretta; local.

R. CAROLINA, L.

Common.

R. HUMILIS, Marsh.

Common. Kelley's and Put-in-Bay the only Islands.

R. *rubiginosa*, L. Sweetbrier. Eglantine.

Frequent. Islands.

R. SETIGERA, Michx. Climbing or Prairie Rose.

Perkins, Groton, Cedar Point, Johnson's Island, Peninsula, Mouse Island, Kelley's Island, Middle Bass; common. Well worth cultivating.

RUBUS, L. Bramble.

R. CANADENSIS, L. Low Blackberry, Dewberry.

Common.

R. HISPIDUS, L. Running Swamp Blackberry.

East of Milan, Berlin, Vermillion, Joseph Smith's, Perkins; local.

R. OCCIDENTALIS, L. Black Raspberry. Thimbleberry.

Common.

R. ODORATUS, L. Purple-flowering Raspberry.

"Near Vermillion River north of Birmingham" Mrs. W. H. Olds. I have seen this handsome species at Buffalo, Ashtabula, Cleveland and in Lorain County within a few rods of Erie County, but no farther west.

R. SETOSUS, Bigel.* Bristly Blackberry.

Prairie, Oxford and Perkins; common.

R. STRIGOSUS, Michx. Wild Red Raspberry.

Old huckleberry swamp near Axtell; rare. "Other places"?

R. TRIFLORUS, Richardson. Dwarf Raspberry.

German settlement, Perkins, and east fork of Vermillion River; rare. Also in the forest west of Castalia, in Sandusky County.

R. VILLOSUS, Ait.   High Blackberry.
　Common.

SPIRÆA, L. Meadow-Sweet.

S. LOBATA, Jacq.*   Queen of the Prairie.
　Southwest of Castalia; local.
　A beautiful plant.

S. SALICIFOLIA, L.   Common Meadow-sweet.
　Oxford, Perkins, Milan, Florence; infrequent.

S. TOMENTOSA, L.   Hardhack.   Steeple-Bush.
　Oxford prairie; very rare.

POMACEÆ.

AMELANCHIER, Medic. June-berry.

A. CANADENSIS, Torr & Gray.   Shad-bush.   Service-
berry.
　Frequent.   Islands.

A. OBLONGIFOLIA, Torr & Gray.
　Cedar Point, Mouse Island, Kelley's Island;
　scarce.

CRATÆGUS, L. Thorn.

C. COCCINEA, L.
　Common.   Put-in-bay; scarce.   North Bass.   No
　other islands.

C. CRUS-GALLI. L.   Cockspur Thorn.
　Frequent.

C. *oxyacantha*, L.   English Hawthorn.
　In a thicket, Vermillion and two places in Huron.
　Seed probably dropped by birds.

C. PUNCTATA, Jacq.
　Perkins, Shinrock, Florence.   Frequent   in
　Florence.   "Marblehead" Gertrude Johnson.

C. SUBVILLOSA, T. & G.   (C. COCCINEA MOLLIS, T. & G.)
　Common.   Kelley's the only Island.

C. TOMENTOSA, L.
　Infrequent.   Kelley's Island.   Middle Bass.

PYRUS, L.

P. AMERICANA, D C.* American Mountain-Ash.
In thickets, Rattlesnake Island, Put-in-Bay and several places in Erie County. Doubtless from seeds dropped by birds.

P. ANGUSTIFOLIA, Ait.*
"Margaretta" Flossie Nolan. Perkins, scarce.

P. ARBUTIFOLIA, L. f. Choke-berry.
Tisdell's, Vermillion; rare.

P. ARBUTIFOLIA MELANOCARPA, Hook.
Milan, Berlin, Vermillion, Marblehead; infrequent

P. *communis*, L. Pear.
In woods or by roadsides, Perkins, Groton, Catawba, Put-in-Bay; rare. "Kelley's Island."

P. CORONARIA. L. American Crab-Apple.
Frequent. Put-in-Bay.

P. *malus*. L. Apple.
Frequent. Islands.

DRUPACEÆ.

PRUNUS, L.

P. AMERICANA, Marshall. Wild Yellow or Red Plum.
Rather frequent. Kelley's Island. Put-in-Bay.

P. *avium*, L. Sweet Cherry.
In several woods where, doubtless, it has started from pits dropped by birds. Kelley's Island.

P. CUNEATA, Raf.*
Oxford prairie; rare.

P. *persica*, Stokes. Peach.
Roadsides; infrequent. Islands. 300,000 bushels of peaches, raised on Catawba, were shipped from there in 1898, enough to have supplied more than a peck to every family in the western half of the United States.

P. SEROTINA, Ehrh. Wild Black Cherry.
> Common. Timber found in the submerged forest, Huron marsh. Mr. W. H. Todd says that these cherries are more attractive to birds than grapes, and that it pays to plant the trees near vineyards for this reason. Are they not worth planting for the timber?

P. VIRGINIANA, L. Choke-Cherry.
> Abundant on Cedar Point and Islands. Much less common elsewhere.

CÆSALPINACEÆ.

CASSIA, L. Senna.

B. CHAMÆCRISTA, L. Partridge Pea.
> Common on the shale in Oxford, Perkins, and Huron near the "slate" cut. Infrequent along railroads in Sandusky. Catawba.

C. MARYLANDICA, L. Wild Senna.
> Margaretta, Johnson's Island, Marblehead; infrequent. "Port Clinton."

CERCIS, L. Judas-tree.

C. CANADENSIS, L. Red-bud.
> Peninsula; frequent. Margaretta; infrequent. Milan; scarce.

GLEDITSCHIA, L. Honey-Locust.

G. TRIACANTHOS, L. Three-thorned Acacia. Honey-Locust.
> Common, especially near Sandusky and in Ottawa county. A tree of great expanse stands on Osborn St. near Hayes Ave.

GYMNOCLADUS, Lam. Kentucky Coffee-tree.

G. CANADENSIS, Lam.
> Distribution peculiar and the tree not generally known. It grows on all of the eight islands on

which I have collected, yet on Put-in-Bay seems limited to onespot near the south point. Marblehead, one standing by the side of the principal street; Catawba; Port Clinton where Dr. Hitchcock said there were fifty on one acre, Margaretta, several places; Perkins, Gurley's; Huron, one by the Sandusky road; Berlin, formerly on Sterling Hill's place and elsewhere; Vermillion, near Axtel; Florence, near Terryville.

PAPILIONACEÆ

AMPHICARPÆA, Ell. Hog Pea-nut.

A. MONOICA, Ell.
Common.

A. PITCHERI, Torr & Gray.*
Perkins, Milan, Cedar Point, Catawba, Islands; frequent.

APIOS, Boerhaave. Ground-nut. Wild Bean.

A. TUBEROSA, Moench.
Rather frequent. "Tubers edible."

ASTRAGALUS, L. Milk-Vetch.

A. CANADENSIS, L.
Shores of the Islands and about Sandusky Bay; frequent.

BAPTISIA, Vent. False Indigo.

B. LEUCANTHA, Torr & Gray.
Oxford and southern Perkins; infrequent.

B. TINCTORIA, R. Br. Wild Indigo.
Oxford, Perkins, eastern Milan, Vermillion, Florence; infrequent.

DESMODIUM, Desv. Tick-Trefoil.

D. ACUMINATUM, DC.
Common. Not on the Islands. Some specimens show a reversion of loments to leaves. See sixth annual report, page 32.

D. CANADENSE, DC.
  Frequent.
D. CANESCENS, DC.
  Common.
D. CILIARE, DC.
  Margaretta Ridge, Berlin Heights, east of Milan
  and Joseph Smith's woods, Perkins; infrequent.
D. CUSPIDATUM, Hooker.
  Infrequent.
D. DILLENII, Darlingt.
  Frequent. Put-in-Bay.
D. ILLINOENSE, Gray.*
  Marblehead, Margaretta, southern Perkins;
  scarce.
D. LINEATUM, DC.*
  Joseph Smith's woods, Perkins; local.
D. MARYLANDICUM, F. Boott.
  Margaretta Ridge; rare.
D. NUDIFLORUM, DC.
  Frequent.
D. PANICULATUM, DC.
  Frequent. Put-in.Bay.
D. RIGIDUM, DC.
  Infrequent.
D. ROTUNDIFOLIUM DC.
  Rather frequent in sandy woods, occurring in, at
  least, fourteen places in Erie County and on the
  Peninsula.
D. SESSILIFOLIUM, Torr. and Gray.*
  Sandy fields on Margaretta Ridge; common.
  Sandhill cemetery. Also ten miles west of Toledo.

LATHYRUS, L. Vetchling.

L. MYRTIFOLIUS, Muhl.
  Huron River near Enterprise. "L. S. & M. S. Ry.
  Sandusky," Elmer Unchrich.
L. OCHROLEUCUS, Hook.
  Peninsula and Islands.

L. PALUSTRIS, L.
Common.

L. VENOSUS, Muhl.*
Margaretta Ridge; considerable.

LESPEDEZA, Michx. Bush-Clover.

L. CAPITATA, Michx.
Common, at least in sandy soil.
Not on the Islands.

L. NUTTALLII, Darl.*
Margaretta Ridge.

L. POLYSTACHYA, Michx.
Margaretta Ridge, East of Milan, Berlin Heights,
Vermillion, Florence; frequent.

L. PROCUMBENS, Michx.
Vermillion; rare.

L. RETICULATA, Pers.
Margaretta, Huron, Marblehead, Catawba.

L. STUVEI INTERMEDIA, Watson.
Frequent.

L. VIOLACEA, Pers.
Frequent.

LUPINUS, L. Lupine.

L. PERENNIS, L. Wild Lupine.
Margaretta Ridge; Joseph's Smith's, Perkins;
east of Milan; local. "Scott's cemetery"
Gertrude Taylor.

MEDICAGO, L. Medick.

M. *lupulina*, L. Black Medick. Nonesuch.
Frequent. Islands.

M. *sativa*, L. Lucerne. Alfalfa.
Sandusky, Perkins, Marblehead, Put-in-Bay;
roadsides, scarce. Can be raised in the dry soil of
the Peninsula and Islands.

MELILOTUS, Juss. Melilot. Sweet Clover.

M. *alba* Desv. White Melilot.
Abundant.

M. *officinalis*, Lam. Yellow Melilot.
Sandusky, Johnson's Island, Put-in-Bay; infrequent.

PHASEOLUS, L.

P. DIVERSIFOLIUS, Pers. (STROPHOSTYLES ANGULOSA, Ell.) Trailing Wild Bean.
Common on sandy shores. Islands.

PSORALEA, L.

P. MELILOTOIDES, Michx.*
Bloomingville cemetery and southeast of Kimball; indigenous but rare.

ROBINIA, L. Locust-tree.

R. PSEUDACACIA, L. Common Locust. False Acacia.
Infrequent. Islands. Naturalized on banks of Huron River and elsewhere.
The first tree of this species taken to Europe, 1638, was still standing in the *Jardin des Plantes*, Paris, in 1890.

TEPHROSIA, Pers. Hoary Pea.

T. VIRGINIANA, Pers. Goat's Rue. Cat-gut.
Castalia cemetery.

TRIFOLIUM, L. Clover.

T. HYBRIDUM, L. Alsike Clover.
Frequent. Put-in-Bay.
T. *pratense*, L. Red Clover.
Common.
T. REFLEXUM, L.* Buffalo Clover.
"Johnson's Island." Minnie Matern.
T. REPENS, L. White Clover.
Common.

VICIA, L. Vetch.

V. AMERICANA, Muhl.
Sandusky, especially along L. S. & M. S. R. R. west of Hancock St., Margaretta Ridge, Catawba, Kelley's Island, North Bass; local.

V. CAROLINIANA, Walt.
Islands, Peninsula and western part of Erie
county; common.
V. *sativa*, L.
Lakeside, North Bass, Rattlesnake Island; rare.

GERANIACEÆ.

ERODIUM, L'Her. Storksbill.

E. *cicutarium*, L'Her.
"East of Milan." Will Bittner.

GERANIUM, L. Cranesbill.

G. CAROLINIANUM, L.
Frequent in cultivated ground. Islands.
G. MACULATUM, L. Wild Cranesbill.
Common. Kelley's the only Island.
G. ROBERTIANUM, L. Herb Robert.
Common in rocky woods on the Peninsula and all
the Islands. In sand, Cedar Point; frequent.
Florence, but scarce so far from the Lake.
Seldom if ever seen in the interior of Ohio or Michigan. I have seen it in Great Britain, where it is
also native but not so common as on our Islands
and Peninsula. Here it probably thrives better
than anywhere farther south in America. It
blooms from May till late in October and adds
much to the beauty of woodland and rocky
shores.

OXALIDACEÆ.

OXALIS, L. Wood-Sorrel.

O. CYMOSA, Small.
Common.

O. STRICTA, L.
Common.
O. VIOLACEA, L.   Violet Wood-Sorrel.
Frequent along a stream in south-eastern Milan
and in woods in southern Perkins. Infrequent in
Berlin, Huron, near the Soldiers' Home and near
the West Harbor. "Florence."

LINACEÆ.

LINUM, L. Flax.

L. SULCATUM.
Widder's woods and Castalia cemetery, Marga-
retta; Sandhill cemetery; Latham's, Catawba;
rare.
L. *usitatissimum*, L.   Common Flax.
Along railroads; infrequent. Kelley's Island.
L. VIRGINIANUM, L.
Dry unbroken ground, especially at the top of
high steep banks, Oxford and east; scarce.

RUTACEÆ.

PTELEA, L. Hop-tree.

P. TRIFOLIATA, L. Shrubby Trefoil.
Common on the Islands and generally on sandy
shores of the Lake. Occurs also in Florence and
Margaretta. One on Cedar point has a circumference of thirty-four inches, one foot above the
ground.

ZANTHOXYLUM, L. Prickly Ash.

Z, AMERICANUM, Mill. Prickly Ash. Toothache-tree.
Perkins, Groton, Cedar Point, Marblehead, Port
Clinton, Kelley's Island, Middle Bass; frequent.

SIMARUBACEÆ.

AILANTHUS, Desf. Tree-of-Heaven.

A. *glandulosa*, Desf. Chinese Sumach.
Naturalized on Cedar Point and in many places in Sandusky, especially about lumber yards and near buildings where the shelter from wind, the reflected sunlight and the protection afforded by the Bay from untimely frosts enable it to thrive better than in most places so far north. Woods, Florence, and creek valleys, Berlin; rare.

POLYGALACEÆ.

POLYGALA, L. Milkwort.

P. SANGUINEA, L.
Abundant on the shale, Oxford and southern Perkins. Huron, south-east of Milan, Berlin, Vermillion; locally common.

P. SENEGA, L.
Margaretta Ridge, Marblehead, Perkins cemetery; scarce. The variety *latifolia* grows at Catawba.

P. VERTICILLATA, L.
Dry soil, especially at the top of steep banks; infrequent.

P. VERTICILLATA AMBIGUA, Wats & Coult.
South of Huron; rare.

EUPHORBIACEÆ.

ACALYPHA, L. Three-seeded Mercury.

A. VIRGINICA, L.
Abundant.

EUPHORBIA, L. Spurge.

E. COMMUTATA, Engelm.
Marblehead, Johnson's Island, Cedar Point,
Willow Point; rare except near the railroad on
Marblehead.

E. COROLLATA, L.
Frequent.

E. *cyparissias*, L. Cypress Spurge.
Spreading in and from cemeteries and yards.
Islands.

E. DENTATA, Michx.* '
Islands, Peninsula and mainland near Sandusky
Bay; frequent.

E. HIRSUTA, Wiegand.*
Common, but not on the Islands.

E. MACULATA, L.
Abundant.

E. MARGINATA, Pursh.
Naturalized in flower gardens, frequent; elsewhere rare.

E. *peplus*, L.*
Along fence, Jefferson St., near Fulton St., Sandusky, where it has been for a number of years.

E. POLYGONIFOLIA, L.
Abundant on sandy shores of Lake Erie. Islands.

E. PRESLII, Guss.
Common.

E. SERPENS, HBK.*
Johnson's Island; rare. A lot in Sandusky, vacant in 1896, but since used for a building site.

CALLITRICHACEÆ.

CALLITRICHE, L. Water-Starwort.

C. HETEROPHYLLA, Pursh.
Berlin; rare.

C. VERNA, L.
Birmingham and Kimball; rare.

LIMNANTHACEÆ.

FLŒRKEA, Willd. False Mermaid.

F. PROSERPINACOIDES, Willd.
Common in alluvial soil.

ANACARDIACEÆ.

RHUS, L. Sumach.

R. AROMATICA, Ait. Fragrant Sumac.
Cedar Point and Marblehead; common. Other parts of the Peninsula, Islands, Margaretta, western Perkins; frequent.

R. COPALLINA, L. Dwarf Sumac.
Oxford and southern Perkins; common. Southeast of Milan.

R. GLABRA, L. Smooth Sumac.
Common.

R. RADICANS, L. (R. TOXICODENDRON,) Poison Ivy.
Everywhere except on Green Island. Common. Berries eaten and seeds distributed by birds.

R. TYPHINA, L. Staghorn Sumac.
Islands, Peninsula and Cedar Point; abundant. Lester Carpenter of Kelley's Island has bookshelves of this wood, and says that one tree was sixteen inches in diameter near the ground, and about fourteen inches, at a height of six feet. Where else does sumac attain such a size?

R. VENENATA, DC. Poison Sumac.
Vermillion; almost exterminated. "Formerly in old huckleberry swamp near Axtel" A. A. Blair and L. W. Washburn.

ILICACEÆ.

ILEX, L. Holly.

I. VERTICILLATA, Gray.   Winterberry.   Black Alder.
Rather frequent.   Green Island.

CELASTRACEÆ.

CELASTRUS, L. Shrubby Bitter sweet.

C. SCANDENS, L.   Wax-work, Climbing Bitter-sweet.
Common.

EUONYMUS, L. Spindle-tree.

E. ATROPURPUREUS, Jacq.   Burning-Bush.   Wahoo.
Frequent.   Kelley's Island.
E. OBOVATUS, Nutt.   Running Strawberry Bush.
Islands; Sugar Rock, Catawba; Hartshorn's;
frequent.   Vermillion River, Florence.

STAPHYLEACEÆ.

STAPHYLEA, L. Bladder-nut.

S. TRIFOLIA, L.   American Bladder-nut.
Frequent.   Green Island.

ACERACEÆ.

ACER, L. Maple.

A. DASYCARPUM, Ehrh.   White or Silver Maple.
Common.   Planted for shade.
Wood used in Sandusky in making baskets.
A. RUBRUM, L.   Red or Swamp Maple.
River banks; infrequent.

A. SACCHARINUM, Wang. Sugar or Rock Maple.

Common in Florence, where there are many sugar bushes. Less common in other parts of the county, on the Peninsula and all the Islands. Wood used by the Sandusky Furniture Company for making bowling alleys, and by the Tool Company for the jaws of hand-screws.

A. SACCHARINUM NIGRUM, Torr & Gray. Black Sugar Maple.

Frequent. Kelley's Island. North Bass.

NEGUNDO, Moench. Ash-leaved Maple. Box Elder.

N. ACEROIDES, Moench.

Vermillion River, Huron River, Pipe Creek, Shinrock, Bay Bridge, Port Clinton, Put-in-Bay; scarce except along rivers.

HIPPOCASTANACEÆ.

AESCULUS, L.

Æ. GLABRA, Willd. Fetid or Ohio Buckeye.

Frequent along streams and on Johnson's Island. Marblehead, Kelley's Island; scarce. Middle Bass, one. "North Bass, one." "Buckeye Island, formerly."

BALSAMINACEÆ.

IMPATIENS, L. Balsam. Jewel-weed.

I. AUREA, Muhl. (I. PALLIDA, Nutt.) Pale Touch-me-not.

Frequent in rich soil in damp woods.
Rattlesnake Island.

I. BIFLORA, Walt. (I. FULVA, Nutt.) Spotted Touch-me-not.

Common, especially on Cedar Point and shores of the Islands.

RHAMNACEÆ.

CEANOTHUS, L. Red-root.

C. AMERICANUS, L. New Jersey Tea.
> Peninsula, Margeretta Ridge, Perkins, Oxford, east of Milan; frequent.

C. OVATUS, Desf.
> Peninsula; frequent.

VITACEÆ.

VITIS, L. Grape.

V. BICOLOR, LeConte. Blue or Winter Grape.
> Infrequent. A vine in Peter Mainzer's woods, German Settlement, Perkins, is about 80 feet high and measures 28¼ inches in circumference.

V. CORDIFOLIA, Michx. Frost or Chicken Grape.
> Milan, Berlin, Vermillion; rather frequent. Johnson's Island.

V. HEDERACEA, Ehrh. (AMPELOPSIS QUINQUEFOLIA, Michx.) Virginia Creeper.
> Common.

V. LABRUSCA, L. Northern Fox Grape.
> Vermillion, Florence, Berlin, Milan, Oxford. Rather frequent in Florence.

V. RIPARIA, Michx. Riverside or Sweet scented Grape.
> Common. Abundant on Cedar Point. Nearly all the wild grape vines near Sandusky and on the Islands and Peninsula are of this species. Wild grapes formerly abounded on the Islands. Vineyards have for many years occupied half or more of the cultivated ground of the Islands,—more than half the entire area of Middle Bass and North Bass. Of late they have been to some extent supplanted by peach orchards. The yield continues good,—between six and nine million pounds annually for Ottawa county, surpassed the last few years by Lake and Cuyahoga counties,—but the price has been low.

TILIACEÆ.

TILIA, L. Linden.

T. AMERICANA, L. Basswood.

Common. Wood used in Sandusky for making excelsior and small boxes. Crayon made in Sandusky is used in nearly every school-house in the United States and to some extent in Europe. For the crayon boxes, basswood logs four feet long, steamed and stripped of bark, are revolved in front of a knife that peels off long sheets of the required thickness. The cores of the logs, about six inches thick, are sent to Muncie, Indiana, for making paper pulp.

MALVACEÆ.

ABUTILON, Gaertn. Indian Mallow.

A. *avicennae*, Gaertn. Velvet-Leaf.

Common. Cultivated in western China for its fibre: here a garden weed.

ALTHÆA.

A. *rosea*, Cav. Hollyhock.

Escaped into streets and vacant lots in a hundred places, in Sandusky; also in many other places in Erie county and on the Islands and Peninsula.

HIBISCUS, L. Rose-Mallow.

H. MOSCHEUTOS, L. Swamp Rose-Mallow.

In marshes connected with Sandusky Bay and the Harbors; frequent. Port Clinton. North Bass. A showy plant.

H. *trionum*, L.　Bladder Ketmia.　Flower-of-an-Hour.
Venice Mallow.　Black-eyed Susan.
Frequent.　Not yet well known, but occurring
throughout Frie county, on the Peninsula and on
Kelley's Island.　Plentiful in some places.

MALVA, L.　Mallow.

M. *moschata*, L.　Musk Mallow.
Scarce.　Kelley's Island.
M. *rotundifolia*, L.　Common Mallow.
Abundant.
M. *sylvestris*, L.　High Mallow.
Rare.

SIDA, L.

S. *spinosa*, L.
Sandusky, Perkins, Peninsula ; local.　Kelley's
Island ; frequent.

HYPERICACEÆ.

HYPERICUM, L.　St John's-wort.

H. ASCYRON, L.　Great St. John's-wort.
Vermillion　River,　Huron　River,　Shinrock ;
infrequent.
H. CANADENSE, L.*
South-east of Milan ; rare.
H. CANADENSE MAJUS, Gray.*
Perkins, Groton ; infrequent.
H. GYMNANTHUM, Engelm & Gray.*
Prairie, Oxford and Perkins ; common.
H. KALMIANUM, L.
Prairie north· and west of Castalia ; common.
Middle Bass ; rare.　" Put-in-Bay."
H. MACULATUM, Walt.
Frequent.　Rattlesnake Island.

H. MUTILUM, L.
Frequent. Common on Oxford prairie.

H. *perforatum*, L. Common St. John's-wort.
Frequent. Common in parts of Berlin. Kelley's Island. Middle Bass.

H. SAROTHRA, Michx. (H. NUDICAULE Walt.)
Orange-grass. Pine-weed.
Oxford; common on the shale. Huron, Vermillion; local.

H. VIRGINICUM, L. (ELODES CAMPANULATA, Pursh.)
Marsh St Johns-wort.
Infrequent.

CISTACEÆ.

HELIANTHEMUM, Pers. Frost-weed.

H. CANADENSE, Michx.
Margaretta Ridge and Perkins; rare.

H. MAJUS, (L) B. S. P.
East of Milan; infrequent. Cedar Point and southern Perkins; local.

LECHEA, Kalm. Pinweed.

L. LEGGETTII, Britt & Holl.
Leonard's Hazel Patch, Perkins.

L. MAJOR, Michx.
Wintergreen woods east of Milan, Bloomingville cemetery, Castalia cemetery, Smith's, Perkins; local. "Cedar Point" Claassen.

L. MINOR, L. (L. THYMIFOLIA of Gray's Manual.)
Vermillion, southern Perkins and east of Milan; local and scarcer than the last.

VIOLACEÆ.

IONIDIUM, Vent.

I. CONCOLOR, Benth & Hook. Green Violet.
Vermillion River, Florence; rare.

VIOLA, L. Violet.

V. BLANDA, Willd. Sweet White Violet.
> One wet field in Margaretta, since plowed up.
> "Perkins." "Berlin."

V. BLANDA PALUSTRIFORMIS, Gray.*
> Damp cool rocks, Vermillion River and tributary
> ravines; scarce.

V. CANINA MUHLENBERGII, Gray. Dog Violet.
> Vermillion River near Birmingham; one specimen.
> Also Rocky Ridge, Ottawa county.

V. CUCULLATA, Ait. Common Blue Violet.
> Abundant. In bloom October 8.

V. LANCEOLATA, L. Lance-leaved Violet.
> Oxford and Perkins prairie; rather frequent.
> Vermillion southeast of the village; locally
> plentiful.

V. OVATA, Nutt.*
> Castalia cemetery; rare.

V. PALMATA, L.
> Sandusky, Catawba; scarce.

V. PEDATIFIDA, G. Don.*
> Marblehead; scarce. Margaretta and Perkins
> rare.

V. PUBESCENS, Ait. Downy Yellow Violet.
> Common.

V. PUBESCENS SCABRIUSCULA, Torr & Gray.
> Perkins, Milan. Apparently common: we have
> confounded it with the species.

V. ROSTRATA, Pursh. Long-spurred Violet.
> Florence; frequent. Berlin Heights, but not
> nearer Sandusky.

V. SAGITTATA, Ait. Arrow-leaved Violet.
> Prairie, Oxford and Perkins; common. East of
> Milan. Vermillion. In bloom October 5.

V. STRIATA, Ait. Pale Violet.
> Common along rivers and, locally, elsewhere.

V. TENELLA, Muhl. (Viola *tricolor arvensis* DC.,
perhaps.) Field Pansy. Cedar Point, Johnson's
Island, Marblehead, Catawba. Put-in-Bay; in-
frequent but apparently indigenous. V. *tricolor*
L., Pansy, persists where it has been cultivated.
Three other species grow in Cuyahoga county.
See page 30.

CACTACEÆ.

OPUNTIA, Mill. Prickly Pear.

O. RAFINESQUII, Engelm.*
Cedar Point and one field in Margaretta;
common. Marblehead; scarce.

THYMELÆACEÆ.

DIRCA, L. Leatherwood. Moosewood.

D. PALUSTRIS, L.
One bush on Beecher's flats, Vermillion River,
southern Florence. "Formerly plentiful" there.

ELÆAGNACEÆ.

SHEPHERDIA, Nutt.

S. CANADENSIS, Nutt.
One spot on east fork Vermillion River; rare.
"Cedar Point," W. A. Kellerman.

LYTHRACEÆ.

AMMANNIA, L.

A. COCCINEA, Rottb.*
Presque Isle Point, Peninsula; local.

LYTHRUM, L. Loosestrife.

L. ALATUM, Pursh.
Common, especially on wet prairies.
Put-in-Bay and Middle Bass the only Islands.

NESÆA, Comm, Juss.

N. VERTICILLATA, HBK. (DECODON VERTICILLATUS.
Ell.) Swamp Loosestrife.
Marshes connected with Bay and Lake; common.
Islands.

ROTALA, L.

R. RAMOSIOR, Koehne.
Marblehead; rare. The only spot in northern
Ohio.

MELASTOMACEÆ.

RHEXIA, L. Deer-Grass.   Meadow-Beauty.

R. VIRGINICA, L.*
Southern Perkins and East of Milan; plentiful in
a few places; regarded rare until 1898.

ONAGRACEÆ.

CIRCÆA, L. Enchanter's Nightshade.

C. ALPINA, L.
Florence, mostly on old logs; scarce.
C. LUTETIANA, L.
Common. Put-in-bay the only Island.

EPILOBIUM, L. Willow-herb.

E. ADENOCAULON, Haussk.
Castalia, Vermillion in old quarry, Marblehead,
Kelley's Island, North Bass; infrequent.

E. ANGUSTIFOLIUM, L. Great Willow-herb. Fire-weed.
Infrequent.

E, COLORATUM. Muhl.
Frequent. Kelley's Island. Middle Bass.

E. LINEARE, Muhl.
Castalia and Peninsula; infrequent.

GAURA, L.

G. BIENNIS, L.
Rather frequent.

LUDWIGIA, L. False Loosestrife.

L. ALTERNIFOLIA, L. Seed-box.
Common on the shale. Cedar Point.

L. PALUSTRIS,. Ell. Water Purslane.
Frequent.

L. POLYCARPA, Short & Peter.
Oxford, Perkins, Vermillion; infrequent.

ŒNOTHERA, L. Evening Primrose.

Œ. BIENNIS, L. Common Evening Primrose.
Common.

Œ. FRUTICOSA, L. Sundrops.
Kimball; locally plentiful.

Œ. OAKESIANA, Robbins.*
Sandusky and probably Cedar Point and elsewhere about the Lake. Not distinguished from
Œ. *bennis* until 1898, probably for the reason
that it is not annual, as described. Several years
ago August Guenther, at my suggestion, pulled up
a large number of Œnotheras on Cedar Point
and elsewhere, but failed to find one with an annual root. One or the other species is very common on the shores of the Islands.

Œ. PUMILA, L.
Oxford, southern Perkins, east of Milan, Vermillion; scarce. "Southern Margaretta," Elsie Johns.

Œ. RHOMBIPETALA, Nutt.*
Cedar Point.

HALORAGIDACEÆ.

MYRIOPHYLLUM, L. Water-Milfoil.

M. SPICATUM, L.
Sandusky Bay, East Harbor, Catawba, Put-in-
Bay; common.

PROSERPINACA, L. Mermaid-weed.

P. PALUSTRIS, L.
Perkins, Castalia, Marblehead; in swamps.

ARALIACEÆ.

ARALIA, L.

A. NUDICAULIS, L. Wild Sarsaparilla.
Rather frequent. Green Island, Kelley's Island.

A. QUINQUEFOLIA Decsne & Planch. Ginseng.
A few years ago frequent; now nearly exterminated. The ginseng dug on Put-in-Bay, 1892 and
1893, sold for about $800 at about $3 a pound.

A. RACEMOSA, L. Spikenard.
Frequent on steep banks of streams, and occurs
in several other places.'

A. TRIFOLIA, Decsne & Planch. Dwarf Ginseng.
Ground-nut.
Two places in Florence.

UMBELLIFERÆ.

ARCHANGELIGA, Hoffm.

A. ATROPURPUREA, Hoffm.
Castalia; frequent. Perkins.
A. HIRSUTA, Torr & Gray.
Sandy soil; infrequent.

CARUM, L. Caraway.

C. *carvi*, L.
Infrequent. Islands.

CHÆROPHYLLUM, L.

C. PROCUMBENS, Crantz.
Infrequent. Kelley's Island.

CICUTA, L. Water Hemlock.

C. BULBIFERA, L.
Frequent. Islands.
C. MACULATA, L. Musquash Root.
Frequent. Kelley's Island.

CONIUM, L. Poison Hemlock.

C. *maculatum*, L.
Roadside, Groton; local.

CRYPTOTÆNIA, DC. Honewort.

C. CANADENSIS, DC.
Frequent.

DAUCUS, L. Carrot.

D. *carota*, L.
A weed in some places in the eastern part of Erie
county. Infrequent or scarce in Sandusky and
elsewhere, but, perhaps, spreading from the east.

ERIGENIA, Nutt.   Harbinger-of-Spring.

E. BULBOSA, Nutt.
    Rather frequent near streams.
    Kelley's Island.

ERYNGIUM, L.

E. YUCCÆFOLIUM, Michx.*
    Rattlesnake-Master.   Button Snake-root.
    Southeast of Kimball; plentiful.   Roadside west
    of Union Corners, and roadside at Joseph Smith's,
    Perkins; rare.

FŒNICULUM, Adans,   Fennel.

F. *vulgare*, Mill.   (F. *officinale*, All.)
    Sandusky and Groton; rare.

HERACLEUM, L.   Cow-Parsnip.

H. LANATUM, Michx.
    Perkins, Florence, Port Clinton; infrequent.

HYDROCOTYLE, L.   Water Pennywort.

H. AMERICANA, L.
    Florence; rare.

OSMORRHIZA, Raf.   Sweet Cicely.

O. BREVISTYLIS, DC.
    Common.
O. LONGISTYLIS, DC.
    Common.

PEUCEDANUM, L.

P. *sativum*, Benth & Hook.   Parsnip.
    Common.   Kelley's the only island.
P. TERNATUM, Nutt.   (TIEDEMANNIA RIGIDA, Coult &
    Rose.)   Cowbane.
    Infrequent.

PIMPINELLA, L.

P. INTEGERRIMA, Benth & Hook.
Frequent, especially on rocky hillsides. Kelley's Island, Put-in-Bay.

SANICULA, L. Sanicle. Black Snakeroot.

S. CANADENSIS, L.
Frequent or common. Put-in-Bay group.

S. MARYLANDICA, L.
Frequent or common. Kelley's Island.
The two species of sanicle are so much alike that I have not always attempted to distinguish between them. The U. S. National Museum has a specimen of S. *trifoliata* from Lorain county, and the same might probably be found in Erie county by diligent searching.

SIUM, L. Water Parsnip.

S. CICUTÆFOLIUM, Schrank.
Frequent. Kelley's Island.

THASPIUM, Nutt. Meadow-Parsnip.

T. AUREUM, Nutt.
Sandusky, Margaretta, Marblehead: infrequent. The so-called variety *atropurpureum* in Florence.

T. AUREUM TRIFOLIATUM. Coult & Rose.
Frequent on the Peninsula and in the western part of Erie county. Put-in-Bay.

T. BARBINODE, Nutt.
Margaretta, Peninsula, Islands; frequent. "Cedar Point."

T. BARBINODE, ANGUSTIFOLIUM, Coult & Rose.
Cedar Point, Johnson's Island, Marblehead, Mouse Island; frequent.

ZIZIA, Koch.

Z. AUREA. Koch.
Frequent. Kelley's Island.

CORNACEÆ.

CORNUS, L. Cornel. Dogwood.

C. ALTERNIFOLIA, L. f.
Florence, Catawba ; scarce.

C. AMOMUM, Mill. (C. SERICEA, L.)
Silky Cornel. Kinnikinnik.
Common.

C. ASPERIFOLIA, Michx.
Common.

C. CANDIDISSIMA, Mill. (C. PANICULATA, L'Her.)
Frequent.

C. CIRCINATA, L'Her. Round-leaved Cornel or Dogwood.
Frequent, especially on the Peninsula and along
the Vermillion River. Kelley's Island.

C. FLORIDA, L. Flowering Dogwood.
Common. Kelley's the only Island.

C. STOLONIFERA, Michx, Red-osier Dogwood.
Castalia; rare. Shore of Lake Erie east of
Huron.

NYSSA, L. Tupelo.

N. MULTIFLORA, Wang. (N. SYLVATICA, Marsh.)
Pepperidge. Sour Gum.
Rich soil; infrequent.

PYROLACEÆ.

CHIMAPHILA, Pursh. Pipsissewa.

C, MACULATA, Pursh. Spotted Wintergreen.
Furnace woods, Vermillion.

C. UMBELLATA, Nutt. Prince's Pine.
Cedar Point; east of Milan; Vermillion River,
Florence, rare.

PYROLA, L. Wintergreen.

P, ELLIPTICA, Nutt. Shin-leaf.
Florence, Milan, Perkins, Cedar Point, Marblehead; infrequent.

P. ROTUNDIFOLIA, L.
Florence, Berlin Heights, Milan, Perkins, Margaretta Ridge; infrequent.

MONOTROPACEÆ.

MONOTROPA, L. Indian Pipe.

M. UNIFLORA, L. Corpse-Plant.
Infrequent.

●

ERICACEÆ.

ARCTOSTAPHYLOS, Adans. Bearberry.

A. UVA-URSI, Spreng.*
Cedar Point; frequent. Vermillion River, Vermillion; rare.

EPIGÆA, L. Ground Laurel.

E. REPENS, L. Trailing Arbutus.
Berlin Heights; rare.

GAULTHERIA, L. Aromatic Wintergreen.

G. PROCUMBENS, L. Creeping Wintergreen.
One woods east of Milan; frequent. Berlin
Heights and Vermillion River; rare. Formerly so
plentiful on the banks of the Vermillion River
north of Birmingham that they were known
locally as the "Wintergreen Banks."

VACCINIACEÆ.

GAYLUSSACIA, H. B. K. Huckleberry.

H. RESINOSA, Torr & Gray. Black Huckleberry.
Oxford and east; frequent.

OXYCOCCUS, Hill. Cranberry.

O. MACROCARPUS, Pers. Large or American Cranberry.
Milan; nearly exterminated. "Formerly east of
Berlin Heights and plentiful near Axtel."

VACCINIUM, L. Blueberry.

V. CORYMBOSUM, L. High-bush or Swamp Blueberry.
A few bushes on and near Tisdale's Vermillion,
and in the old swamp near Axtel where years
ago, " grew a thousand bushels of berries." See
page 31.

V. PENNSYLVANICUM, Lam. Dwarf Blueberry.
Vermillion River, Vermillion; rare.

V. VACILLANS, Solander. Low Blueberry.
Frequent from the Huron River east. This and
the Black Huckleberry are the only *Ericaceæ* often
met with in Erie county and these not often west
of the Huron River. I know of none of this order
on the Islands and, excepting the Shin-leaf and
"Indian Pipe," none on the Peninsula.

PRIMULACEÆ.

ANAGALLIS, L. Pimpernel.

A. *arvensis*, L. Common Pimpernel.
"Sandusky." Victor Hommel.

DODECATHEON, L. American Cowslip.

D. MEADIA, L.* Shooting-Star.
Castalia; rare. Called also Pride-of-Ohio, but
probably not one in a thousand of the people now
living in Ohio ever saw it growing wild.

LYSIMACHIA, L. Loosestrife.

L. *nummularia*, L. Moneywort.
Frequent in damp places along roads and occasional elsewhere. Middle Bass.

L. QUADRIFOLIA, L.
Rather frequent.

L. STRICTA, Ait.
Infrequent. Bass Islands.

L. THYRSIFLORA, L. Tufted Loosestrife.
Perkins, Huron, Cedar Point, Catawba; infrequent.

SAMOLUS, L. Water Pimpernel. Brook-weed.

S. VALERANDI AMERICANUS, Gray.
Florence, Shinrock, Huron, Milan, Groton; infrequent.

STEIRONEMA, Raf.

S. CILIATUM, Raf.
Common.

S. LONGIFOLIUM, Gray.
Sandusky, Oxford, Margaretta, Peninsula, Put-in Bay, Middle Bass, Rattlesnake Island; frequent.

OLEACEÆ.

FRAXINUS, L. Ash.

F. AMERICANA, L. White Ash.
Common. Wood used by the Sandusky Tool Company for hoe handles.

F. PUBESCENS, Lam. Red Ash.
Frequent. Islands. On Kelley's Island fruit 2¼ inches long and 5–12 inch wide.

F. QUADRANGULATA, Michx. Blue Ash.
Islands and Peninsula; frequent. Margaretta Ridge.

F. SAMBUCIFOLIA, Lam. Black Ash.
Infrequent. Islands.

F. VIRIDIS, Michx. f. Green Ash.
Cedar Point and Vermillion River.

LIGUSTRUM, L.

L. *vulgare*, L. Privet. Prim.
Cedar Point, Milan, etc; rare.

SYRINGA, L.

S. *vulgaris*, L. Lilac.
Kelley's Island; well established. Sandusky.

GENTIANACEÆ.

BARTONIA, Muhl.

B. TENELLA, Muhl.
East of Milan; rare.

FRASERA, Walt. American Columbo.

F. CAROLINENSIS, Walt.
Margaretta Ridge, Perkins, Huron, Berlin; scarce.

GENTIANA, L. Gentian.

G. ANDREWSII, Griseb. Closed Gentian.
Frequent along ditches.

G. CRINITA, Froel. Fringed Gentian.
Castalia, southern Perkins, eastern Milan, Oxford
near Huron River; infrequent. "Marblehead."

G. DETONSA Rottb. (G. SERRATA, Gunner.)
Vermillion River, Florence; one young plant found
on wet shale cliff.

G. PUBERULA, Michx.*
Southern Perkins; beautiful but very rare.

G. QUINQUEFLORA, Lam.
Vermillion River; frequent on the east fork. Margaretta Ridge; rare. The variety *occidentalis* in southern Perkins.

SABBATIA, Adans.

S. ANGULARIS, Pursh.
"Florence, 1888." Josephine Fish.
Eastern Milan and Vermillion River, Florence; scarce.

APOCYNACEÆ.

APOCYNUM, L.

A. ANDROSÆMIFOLIUM, L. Spreading Dogbane.
Frequent. Put-in-Bay. Middle Bass.
A. CANNABINUM, L. Indian Hemp.
Frequent but on lower ground. Islands.

VINCA, L.

V. *minor* L. Periwinkle. Myrtle.
Spreading in and from yards and cemeteries. Kelley's Island. Middle Bass.

ASCLEPIADACEÆ.

ACERATES, Ell. Green Milkweed.

A. LONGIFOLIA, Ell.*
Prairie; Oxford, Perkins. Huron; frequent.
A. VIRIDIFLORA, Ell.
Oxford, Margaretta, Cedar Point, Marblehead, Catawba. Infrequent, except on Marblehead, where the "variety" *lanceolata* also occurs.

ASCLEPIAS, L. Milkweed.

A. INCARNATA, L. Swamp Milkweed.
  Common.

A. INCARNATA PULCHRA, Pers.
  Castalia; rare.

A. OBTUSIFOLIA, Michx.*
  In sand, Margaretta Ridge, Castalia cemetery,
  southern Perkins; rare.

A. PHYTOLACCOIDES, Pursh. Poke Milkweed.
  In nine places, but scarce. Put-in-Bay.

A. PURPURASCENS, L. Purple Milkweed.
  Perkins, Margaretta, Groton, Marblehead,
  Catawba; infrequent.

A. QUADRIFOLIA, Jacq.
  Huron River and Perkins; rare.

A. SULLIVANTII, Engelm.*
  Oxford and Sandusky; scarce.

A. SYRIACA, L. Common Milkweed or Silkweed.
  Common.

A. TUBEROSA, L. Butterfly-weed. Pleurisy-root.
  Frequent. Put-in-Bay. North Bass.

A. VERTICILLATA, L.
  Southern Margaretta, Groton, Marblehead,
  Catawba; scarce.

CONVOLVULACEÆ.

CONVOLVULUS, L. Bindweed.

C. *arvensis*, L. Small Bindweed.
  Sandusky and Islands; local.

C. SEPIUM, L. (CALYSTEGIA SEPIUM, R. Br.)
  Hedge Bindweed.
  Common. A rank weed in corn fields in Perkins.
  On portions of the bay shore of Cedar Point so
  thick as to make walking difficult.

C. SEPIUM REPENS, Gray.*
>Oxford; frequent? Catawba. "Marblehead,"
>U. G. Sanger.

IPOMŒA, L. Morning Glory.

I. PANDURATA, Meyer. (I. FASTIGIATA. Sweet.) Wild
>Potato-vine. Man-of-the-earth.
>Frequent.

I. *purpurea*, Roth. Morning-glory.
>Escaped into roads and waste places, Sandusky,
>Peninsula, Put-in-Bay, North Bass; infrequent.

CUSCUTACEÆ.

CUSCUTA, L. Dodder.

C. ARVENSIS, Beyrich.*
>Oxford, Florence, Port Clinton; rare.

C. CHLOROCARPA, Engelm.*
>Catawba; frequent. East Harbor, Castalia,
>Willow Point, Sandusky. Oxford; infrequent.

C. DECORA, Engelm.*
>Marblehead; rare.

C. GRONOVII, Willd.
>Common.

C. INFLEXA, Engelm.*
>Oxford and Margaretta Ridge; scarce.

G. TENUIFLORA, Engelm.
>Perkins, Oxford, Port Clinton, Put-in-Bay;
>Infrequent.

POLEMONIACEÆ.

PHLOX, L.

P. DIVARICATA, L.
>Common. A specimen from Johnson's Island has
>narrow, acuminate, corolla lobes.

P. PANICULATA, L.
Spreading from gardens to roadsides in several places.

P. PILOSA, L.
Margaretta Ridge, Oxford, southern Perkins, Huron, Catawba; locally common.

P. SUBULATA, L.   Ground or Moss Pink.
Catawba; frequent. Vermillion , or Florence; rare. "Berlin" Sterling Hill.

POLEMONIUM, L. Greek Valerian.

P. REPTANS, L.
Near the Huron and Vermillion rivers; infrequent, "Hartshorn's, Peninsula." Pearl Green.

HYDROPHYLLACEÆ.

HYDROPHYLLUM, L. Waterleaf.

H. APPENDICULATUM, Michx.
Frequent, especially on the Islands and Peninsula.

H. CANADENSE, L.
Florence and Vermillion; rare.

H. MACROPHYLLUM, Nutt.
One spot on west bank of west fork of Vermillion River; a dozen or more plants growing with a few of the preceding species.   Unknown elsewhere so far north.

H. VIRGINICUM, L.
Common. Islands, except Kelley's and Put-in-Bay.

PHACELIA, Juss.

P. PURSHII, Buckley.
Johnson's Island; common. Milan, Vermillion, Peninsula, Kelley's Island; scarce.

BORAGINACEÆ.

BORAGO, L.

B. *officinalis*, L.   Borage.
Spontaneous near the Soldiers' Home.

CYNOGLOSSUM, L.

C. *officinale*, L.   Hound's-tongue.
Common.

ECHINOSPERMUM, Lehm. Stickseed.

E. *lappula*, Lehm.
Peninsula, Kelley's Island, Middle Bass, Perkins, Sandusky; rather frequent.
E. VIRGINICUM, Lehm.   Beggar's Lice.
Frequent. Kelley's Island. Put-in-Bay.

ECHIUM, L. Viper's Bugloss.

E. *vulgare*, L.   Blue-weed.
Well established in the L. E. & W. freight yard, Sandusky.

LITHOSPERMUM, L.

L. *arvense*, L.   Corn Gromwell.
Abundant   One of the worst weeds on Kelley's Island and elsewhere.
L. CANESCENS, Lehm.   Hoary Puccoon.
Peninsula,   Margaretta,   southern   Perkins; infrequent.
L. HIRTUM, Lehm.*   Hairy Puccoon.
Cedar Point; common.

MERTENSIA, Roth. Lungwort.

M. VIRGINICA, DC.   Virginia Cowslip.   Blue-bells.
Johnson's   Island,   Huron   River;   frequent. Marblehead, Kelley's Island, North Bass, Berlin, Vermillion River; infrequent or scarce.

✦

MYOSOTIS, L. Scorpion-grass.

M. VERNA, Nutt.
Rather frequent. Put-in-Bay.

ONOSMODIUM, Michx.

O. CAROLINIANUM, DC.
Margaretta, western Perkins, Peninsula, Johnson's Island; infrequent.

VERBENACEÆ.

LIPPIA, L.

L. LANCEOLATA, Michx. Fog-fruit.
Sandusky, Margaretta, Groton, Johnson's Island, Peninsula, Put-in-Bay; infrequent.

VERBENA, L. Vervain.

V. ANGUSTIFOLIA, Michx.
Common in dry calcareous soil. Kelley's the only island.

V. BRACTEATA, Lag & Rodr.*
Near the L. E. & W. freight house; rare.

V. HASTATA, L. Blue Vervain.
Common.

V. URTICAEFOLIA, L. White Vervain.
Frequent. Islands. Hybrids between this and the preceding occur.

LABIATÆ.

BLEPHILIA, Raf.

B. CILIATA, Raf.
Johnson's Island, Marblehead, Catawba, Kelley's Island, Put-in-Bay, Margaretta, western Perkins; locally plentiful.

B. HIRSUTA, Benth.
   In woods, Erie county and Catawba; infrequent.

CALAMINTHA. Lam.

C. CLINOPODIUM, Benth. Basil.
   Islands, Peninsula, Cedar Point; common.
   Smith's woods, Perkins.
C. NUTTALLII, Benth.
   Prairies, Castalia and Marblehead; common.

COLLINSONIA, L. Horse Balm.

C. CANADENSIS, L. Rich-weed.
   Frequent.

HEDEOMA, Pers.

H. PULEGIOIDES, Pers. American Pennyroyal.
   Common.

ISANTHUS, Michx.

I. CAERULEUS, Michx. False Pennyroyal.
   Dry calcareous soil; frequent, especially about
   quarries. Kelley's Island. Common on Marblehead.

LAMIUM, L. Dead-Nettle.

I, *amplexicaule*, L.
   Throughout but scarce. Islands.
L. *pnrpureum*, L.
   "Soldiers' Home." Carl Anderson.

LEONURUS, L.

L. *cardiaca*, L. Motherwort.
   Common.

LOPHANTHUS, Benth. Giant Hyssop.

L. NEPETOIDES, Benth.
   Peninsula; frequent. Kelley's Island, Cedar
   Point, Johnson's Island, Groton, Perkins, Bloom-
   ingville, Florence; infrequent.

L. SCROPHULARIAEFOLIUS, Benth.
> East of Milan; rare. Also at Oak Harbor, Ottawa county.

LYCOPUS, L.   Water Hoarhound.

L. RUBELLUS, Moench.
> Infrequent. Islands.

L. SINUATUS, Ell.
> Frequent. Islands.

L. VIRGINICUS, L.   Bugle-weed.
> Common.

MARRUBIUM, L.   Hoarhound.

M. *vulgare*, L.   Common Hoarhound.
> Islands and Peninsula; common. Margaretta Sandusky, Milan; local.

MELISSA, L.   Balm.

M. *officinalis*, L.   Common Balm.
> Woods, Put-in-Bay and Vermillion; rare.

MENTHA, L.   Mint.

M. CANADENSIS, L.   Wild Mint.
> Common.

M. *piperata*, L.   Peppermint.
> Frequent, especially about Castalia. "The continuous inhalation of the oil for several days will cure catarrh."

M. *viridis*, L.   Spearmint.
> Common. Put-in-Bay the only island.

M. CLINOPODIA, L.   MONARDA, L. horse-mint.
> Milan; rare.

M. FISTULOSA, L.   Wild Bergamot.
> Common. The variety *mollis* seems to be the more common form.

NEPETA, L. Cat-Mint,

N. *cataria*, L. Catnip.
Common.

N. *glechoma*, Benth. Ground Ivy. Gill.
Common. Not noticed on the Islands, except Rattlesnake, where it appeared about 1892, and Put-in-Bay. Along rivers it has become superabundant.

PHYSOSTEGIA, Benth. False Dragon-head.

P. VIRGINIANA, Benth.
Marblehead, Put-in-Bay, Middle Bass, Groton, eastern Sandusky; scarce.

PRUNELLA, L. Self-heal.

P. VULGARIS, L. Heal-all.
Common.

PYCNANTHEMUM, Michx. Mountain Mint.

P. LANCEOLATUM, Pursh.
Castalia; common. Oxford, Milan, Peninsula; frequent. Put-in-Bay.

P. LINIFOLIUM, Pursh.
Oxford prairie and Vermillion River flats; rare.

P. MUTICUM PILOSUM, Gray.
East of Port Clinton; rare.

SATUREIA, L. Savory.

S. *hortensis*, L. Summer Savory.
Well established in and near the village of Marblehead.

SCUTELLARIA, L. Skullcap.

S. GALERICULATA, L.
Common. Put-in-Bay and Middle Bass the only islands.

S. LATERIFLORA, L. Mad-dog Skullcap.
Common.
S. NERVOSA, Pursh.
Vermillion, woods east of the river and Florence
along west fork; rare.
S. PARVULA, Michx.
Mostly in calcareous soil, Margaretta, Peninsula,
Kelley's Island; frequent.
S. VERSICOLOR, Nutt.
Marblehead; frequent. Cedar Point, Johnson's
Island, Put-in-Bay, Catawba, Margaretta, Per-
kins; infrequent.

STACHYS, L. Hedge-Nettle.

S. ASPERA, Michx.
Sandusky, Cedar Point, Peninsula; common.
Middle Bass, North Bass.
S. TENUIFOLIA, Willd. (S. ASPERA GLABRA, Gray.)
Old Woman Creek, Berlin Heights; rare.

TEUCRIUM, L. Germander.

T. CANADENSE, L. Wood Sage.
Common especially on the shores of the Islands.

SOLANACEÆ.

DATURA, L. Jamestown or Jimson-weed.

D. *stramonium*, L.
Margaretta; frequent; elsewhere scarce.
D. *tatula*, L.
Frequent. Kelley's Island.

LYCIUM, L. Matrimony Vine.

L. *vulgare*, Dunal.
Escaped from gardens in some places. Kelley's
Island.

LYCOPERSICUM, Hill.

L. *esculentum*, Mill. Tomato.
Sandusky; well established near the Bay. Kelley's Island. Put-in-Bay.

NICANDRA, Adans. Apple of Peru.

N. *physaloides*, Gaertn.
Perkins; scarce.

PHYSALIS, L. Ground Cherry.

P. HETEROPHYLLA, Nees. (P. VIRGINIANA, Gray.)
Common.

P. HETEROPHYLLA AMBIGUA, Gray.
Marblehead.

P. HETEROPHYLLA NYCTAGINEA, Dunal.
Huron, Milan, Perkins, Danbury.

P. LANCEOLATA, Michx.
Sandusky, Perkins, Port Clinton, Kelley's Island,
" Marblehead."

P. PHILADELPHICA, Lam.
Perkins, Groton.

P. PRUINOSA, L.
Kelley's Island.

SOLANUM, L. Nightshade.

S. CAROLINENSE, L. Horse-Nettle.
Several places near railroads; scarce.

S. *dulcamara*, L. Bittersweet.
Frequent, especially on the Peninsula and Islands.
Abundant in Lake woods east of Port Clinton. Appearing to be indigenous.

S. NIGRUM, L. Common Nightshade.
Common.

S. ROSTRATUM, Dunal.
Marblehead, about the quarry, where the dry
soil seems adapted to this western weed, but we
hope Mr. Harsh has succeeded in exterminating
it. Put-in-Bay and "west of Sandusky," 1895.

SCROPHULARIACEÆ.

CASTILLEJA, L. Painted-Cup.

C. COCCINEA, Spreng. Scarlet Painted-Cup.
Hartshorn's, Peninsula and Catawba; rare.

CHELONE, L, Turtle-head.

C. GLABRA, L. Snake-head.
Throughout Erie county; infrequent.

CONOBEA, Aublet.

C. MULTIFIDA, Benth.*
Prairies, Castalia, Marblehead, Kelley's Island;
scarce.

GERARDIA, L.

G. AURICULATA, Michx.*
Marblehead; rare.

G. FLAVA, L. Downy False Foxglove.
"Huron River?" Henry Schoepfle.

G. PURPUREA, L. Purple Gerardia.
Castalia, where it adorns the grounds of the
Trout Club, Oxford, southern Perkins, Perrin's,
Milan, Cedar Point, Peninsula; infrequent.

G. PURPUREA PAUPERCULA, Gray.*
Oxford and southern Perkins; rare.

G. QUERCIFOLIA, Pursh. Smooth False Foxglove.
Infrequent.

G. TENUIFOLIA, Vahl. Slender Gerardia.
Frequent. Kelley's Island.

GRATIOLA, L. Hedge-Hyssop.

G. SPHAEROCARPA, Ell.*
DeLamater's, Kimball; rare.

G. VIRGINIANA, L.
Rather frequent.

ILYSANTHES, Raf.

I. RIPARIA, Raf. False Pimpernel.
Sandusky, Huron River, Peninsula; infrequent.

LINARIA, Juss. Toad Flax.

L. *vulgaris*, Mill. Butter and Eggs.
Common.

MIMULUS, L. Monkey-flower.

M. ALATUS, Solander.
Frequent in the eastern part of Erie county.
Milan and Perkins; infrequent.

M. RINGENS, L.
Frequent. Bass Islands.

PEDICULARIS, L. Lousewort.

P. CANADENSIS, L. Wood Betony.
Infrequent. Kelley's Island. Put-in-Bay.

P. LANCEOLATA, Michx.
Milan, Margaretta, Perkins; infrequent.

PENTSTEMON, Mitchell. Beard-tongue.

P. PUBESCENS, Solander.
Frequent, especially on the Islands and Peninsula.

SCROPHULARIA, L. Figwort.

S. NODOSA MARYLANDICA. Gray.
Frequent. Islands,

SEYMERIA, Pursh.

S. MACROPHYLLA, Nutt. Mullein-Foxglove.
Cedar Point, Port Clinton, Vermillion River;
scarce.

VERBASCUM, L. Mullein.

V. *blattaria*, L. Moth Mullein.
Frequent. Kelley's Island. Middle Bass.

V. *thapsus*, L. Common Mullein.
Common.

VERONICA, L. Speedwell.

V. ANAGALLIS, L. Water Speedwell.
Margaretta, Huron, Berlin, Kelley's Island;
infrequent.

V. *arvensis*, L. Corn Speedwell.
Common.

V. *hederæfolia*, L.* Ivy-leaved Speedwell.
"Yard on east Market St., Sandusky." Ione
Pratt.

V. OFFICINALIS, L. Common Speedwell.
Margaretta Ridge and east of Port Clinton;
rare. "Florence." Josephine Fish.

V. PEREGRINA, L. Neckweed. Purslane Speedwell.
Frequent. Put-in-Bay, North Bass, Rattlesnake
Island.

V. SCUTELLATA, L. Marsh Speedwell.
Infrequent.

V. SERPYLLIFOLIA, L. Thyme-leaved Speedwell.
Frequent. Put-in-Bay.

V. VIRGINICA, L. Culver's-root. Culver's Physic.
Infrequent.

LENTIBULARIACEÆ.

UTRICULARIA, L. Bladderwort.

U. GIBBA, L.*
Cedar Point; local.

U. VULGARIS, L. Greater Bladderwort.
Sandusky Bay and East Harbor; frequent. Castalia; infrequent.

OROBANCHACEÆ.

APHYLLON, Mitchell.

A. UNIFLORUM, Gray. One-flowered Cancer-root.
Sandusky, three places; "Bogart" James D.
Parker, Jr.; Florence; "Catawba" Earl Covell:
scarce.

CONOPHOLIS, Wallroth. Squaw-root. Cancer-root.

C. AMERICANA, Wallroth.
Local. Put-in-Bay, northwest woods; plentiful.
Perkins, big woods. Florence; two places.

EPIFAGUS, Nutt. Beech-drops. Cancer-root.

E. AMERICANUS, Nutt. (EPIPHEGUS VIRGINIANA, Bart).
Florence, Vermillion, Berlin; frequent.

BIGNONIACEÆ·

TECOMA, Juss. Trumpet-flower.

T. RADICANS, Juss. Trumpet Creeper.
Frequent in woods and probably indigenous.
Abundant on Cedar Point. Islands.

ACANTHACEÆ.

DIANTHERA, Gronov. Water-Willow.

D. AMERICANA, L.
Marblehead, Put-in-Bay, Middle Bass; rare.
"Mills Creek; plentiful" Hommel.

PHRYMACEÆ.

PHRYMA, L. Lopseed.

P. LEPTOSTACHYA, L.
Frequent. Kelley's Island. Put-in-Bay.

PLANTAGINACEÆ.

PLANTAGO, L. Plantain.

P. ARISTATA, Michx.
Sandy field on Margaretta Ridge and near L. E.
& W. freight house, Sandusky; rare.

P. CORDATA, Lam.
Huron and Florence; rare.

P. *lanceolata*, L. Ribgrass. Ribwort. English
Plantain.
Frequent but not common in most parts. Kelly's
Island, Put-in-Bay.

P. MAJOR, L. Common Plantain.
Common.

P. RUGELII, Decaisne.
More common than the preceding.

P. VIRGINICA, L.
Sandy field on Margaretta Ridge; rare.

RUBIACEÆ.

CEPHALANTHUS, L. Button-bush.

C. OCCIDENTALIS, L.
Common.

GALIUM, L. Bedstraw. Cleavers.

G. APARINE, L. Cleavers. Goose-Grass.
Abundant.

G. ASPRELLUM, Michx. Rough Bedstraw.
Infrequent. Islands.

G. BOREALE, L. Northern Bedstraw.
Perkins, Margaretta, Marblehead, Catawba,
Kelley's Island; scarce.

G. CIRCÆZANS, Michx. Wild Liquorice.
Rather common. Put-in-Bay, Middle Bass,
Rattlesnake Island.

G. CONCINNUM, Torr & Gray.
> Common. Not on the Islands.

G. LANCEOLATUM, Torr. Wild Liquorice.
> Florence, Vermillion, Berlin Heights; rare.

G. PILOSUM, Ait.
> Frequent. One specimen shows a reversion of flowers to leaves.

G. TRIFIDUM, L. Small Bedstraw.
> Frequent. Put-in-Bay. Middle Bass. The variety *pusillum* occurs at Castalia and "Cedar Point."

G. TRIFIDUM LATIFOLIUM, Torr.
> Infrequent.

G. TRIFLORUM, Michx. Sweet-scented Bedstraw.
> Frequent. Rattlesnake Island.

HOUSTONIA, L.

H. CÆRULEA, L. Bluets. Innocence.
> Not found near Sandusky but in many places from southern Perkins south and east. East of Milan I have seen several million blossoms on three or four acres of ground, appearing at a distance as if a light snow had fallen, not completely covering the grass.

H. CILIOLATA, Torr.
> Marblehead; common. Margaretta. Soldier's Home.

H. LONGIFOLIA, Gaertn.
> Rocky shores of Rattlesnake Island and Put-in-Bay; frequent. Marblehead.

MITCHELLA, L. Partridge-berry.

M. REPENS, L.
> Banks of Vermillion River and tributaries; common. Old Woman Creek at Berlin Heights; frequent. Milan, Perkins, Groton; scarce.

CAPRIFOLIACEÆ.

LONICERA, L.　Honeysuckle.

L. GLAUCA, Hill.
Margaretta Ridge; rare.

L. GLAUCESCENS, Rydb.
Infrequent. Islands.

L. SEMPERVIRENS, L. Trumpet or Coral Honeysuckle.
Woods near Huron, where the seed was doubtless dropped by birds; rare.

SAMBUCUS. L.　Elder.

S. CANADENSIS, L. Common Elder.
Common.

S. RACEMOSA, L. Red-berried Elder.
Eastern Sandusky; east of Milan; Vermillion River, Florence; scarce.

SYMPHORICARPOS, Juss.　Snowberry.

S. ORBICULATUS, Moench. (S. VULGARIS, Michx.)
Indian Currant. Coral-berry.
Sandusky and Milan; escaped.

S. RACEMOSUS, Michx. Snowberry.
Marblehead; common. Elsewhere scarce.

S. RACEMOSUS PAUCIFLORUS, Robbins.
Cedar Point; common-

TRIOSTEUM, L.　Horse-Gentian.

T. PERFOLIATUM, L. Fever-wort.
Frequent.

VIBURNUM, L.　Arrow-wood.

V. ACERIFOLIUM, L. Dockmackie.
Frequent from the Huron River east. Put-in-Bay.

V. DENTATUM, L.
Florence and eastern Berlin; infrequent.

V. LENTAGO, L. Sweet Viburnum. Sheep-berry.
Infrequent. Kelley's Island, Middle Bass.

V. OPULUS, L. Cranberry-tree.
"Groton" and big woods, Perkins; rare.

V. PUBESCENS, Pursh.
Marblehead, Catawba, Kelley's Island, Put-in-Bay; frequent.

VALERIANACEÆ.

VALERIANA, L. Valerian.

V. PAUCIFLORA, Michx.
Lake woods east of Port Clinton, Florence, Milan; rare.

VALERIANELLA, Poll. Corn-Salad. Lamb-Lettuce.

V. *olitoria*, Poll.
Shinrock; rare.

V. RADIATA, Dufr.
Perkins, Milan, Shinrock; scarce.

V. WOODSIANA, Walp.*
Woodbury's woods, Berlin; local.

DIPSACACEÆ.

DIPSACUS, L. Teasel.

D. *sylvestris*, Mill.
Common. Kelley's the only island.

CUCURBITACEÆ.

ECHINOCYSTIS, Torr & Gray. Wild Balsam-apple.

E. LOBATA, Torr & Gray.
Lake woods east of Port Clinton; abundant. Elsewhere infrequent.

SICYOS, L.

S. ANGULATUS, L.   One-seeded Bur-Cucumber.
Green Island; common.   Rattlesnake Island, Put-
in-Bay, Catawba, Port Clinton, Cedar Point,
Sandusky; infrequent.

CAMPANULACEÆ.

CAMPANULA, L.   Bellflower.

C. AMERICANA, L.   Tall Bellflower.
Common.

C. APARINOIDES, Pursh.   Marsh Bellflower.
Cedar Point, Venice, Peninsula; locally common.

C. ROTUNDIFOLIA.   Harebell.
Common on rocky shores but apparently absent
from Kelley's Island.

LOBELIA, L.

L. CARDINALIS, L.   Cardinal-flower.
Infrequent.   Islands.

L. INFLATA, L.   Indian Tobacco.
Rather frequent.   Put-in-Bay.

L. KALMII, L.
Common on rocky shores.   Florence; rare.

L. SPICATA, Lam.
Common on the prairies.

L. SYPHILITICA, L.   Great Lobelia.
Common.   Kelley's, Middle Bass and North Bass
the only islands.

SPECULARIA, Heister.,   Venus's Looking-glass.

S. PERFOLIATA, A. DC.
Infrequent.   Kelley's Island, Put-in-Bay.

CICHORIACEÆ.

CICHORIUM, L. Chicory. Succory.

C. *intybus*, L.

Roadsides in a number of places; local. Common at Port Clinton and Catawba. Kelley's Island, Middle Bass.

HIERACIUM, L. Hawkweed.

H. CANADENSE, Michx.*

Huron, Milan, Oxford, Marblehead, Catawba; infrequent.

H. GRONOVII, L. Hairy Hawkweed.

Infrequent. The "variety" *subnudum* in the Bloomingville cemetery.

H. PANICULATUM, L.

Vermillion River and Berlin Heights; infrequent.

H. SCABRUM, Michx.

Frequent.

KRIGIA, Schreb. Dwarf Dandelion.

K. AMPLEXICAULIS, Nutt.

Frequent in Milan Township. Elsewhere infrequent. Kelley's Island.

LACTUCA, L. Lettuce.

L. ACUMINATA, Spreng.

Perkin's, Margaretta, Port Clinton; infrequent.

L. ALPINA, Benth & Hook, (L. LEUCOPHÆA, Gray.)

Frequent. Kelley's Island, Put-in-Bay.

L. CANADENSIS, L. Wild Lettuce.

Common.

L. FLORIDANA, Gaertn.

Margaretta Ridge, Cedar Point, Peninsula, Put-in-Bay, Green Island; frequent.

L. *scariola*, L. Prickly Lettuce.

Abundant. One of the worst weeds.

PRENANTHES, L. Rattlesnake-root.

P. ALBA, L. White-lettuce.
Common.

P. ALTISSIMA, L.
Infrequent. Put-in-Bay.

P. ASPERA, Michx.*
Prairie east of Kimball; rare.

P. CREPIDINEA, Michx.
Near Pipe Creek in German Settlement woods;
rare.

P. RACEMOSA, Michx.
Prairies. West of Castalia; frequent. Oxford,
Groton, "Perkins," "Gypsum"; infrequent or
scarce.

SONCHUS, L. Sow-Thistle.

S. *asper*, Vill. Spiny-leaved Sow-thistle.
Infrequent. Islands.

S. *oleraceus*, L. Common Sow-Thistle.
Frequent. Islands.

TARAXICUM, L. Dandelion.

T. *officinale*, Weber. Common Dandelion.
Abundant. "In blossom when the boys were
skating" Freyensee.

TRAGOPOGON, Goats-beard.

T. *porrifolius*, L. Salsify. Oyster-plant.
Roadsides; infrequent. Islands.

T. pratensis, L. Goats-beard.
Sandusky, in vacant lots near Central Avenue
and elsewhere; spreading.

COMPOSITAE.

ACHILLEA, L. Yarrow.

A. MILLEFOLIUM, L. Common Yarrow or Milfoil.
Abundant.

ACTINELLA, Nutt.

A. ACAULIS GLABRA, Gray.*
Marblehead prairie; infrequent but occurring at
places widely separated and, apparently,
indigenous.

ACTINOMERIS, Nutt.

A. SQUARROSA, Nutt.
Frequent on flood grounds of streams.

AMBROSIA, L. Ragweed.

A. ARTEMISIÆFOLIA, L. Ragweed. Roman Worm-
wood.
Abundant. After *Setaria glauca* probably the
worst weed.
A. TRIFIDA, L. Great Ragweed.
Common. The so-called variety *integrifolia* is
infrequent.

ANTENNARIA, Gaertn. Everlasting.

A. PLANTAGINEA, R. Br. Plantain-leaved Everlasting.
Common. Kelley's and Put-in-Bay the only islands.
A specimen collected on Marblehead by Ralph H.
McKelvey is what Greene would call *A. neglecta*
and one in Perkins by Will Sprow *A. neodioica*.

ANTHEMIS, L. Chamomile.

A. *cotula*, L. May-weed.
Common.

ARCTIUM, L. Burdock.

A. *lappa majus*, Gray.
"Bogart" H. D. Banks.

A. *lappa minus*, Gray.
Common.

ARTEMISIA, L. Wormwood.

A. *annua*, L.
Sandusky, well established near the Big Four docks.

A. BIENNIS, Willd.
Sandusky, Castalia, Johnson's Island, Marblehead, Middle Bass, North Bass; frequent only near railroads or docks.

A. CAUDATA, Michx.*
Cedar Point and Marblehead sand spit; common.

A. LUDOVICIANA, Nutt.* Western Mugwort.
Established in one spot on embankment of L. S. & M. S. Ry., eastern Sandusky.

A. *vulgaris*, L. Common Mugwort.
Escaped in cemeteries and from gardens to roads; scarce.

ASTER, L.

A. AZUREUS, Lindl.
Sandy soil from Margaretta Ridge to Berlinville; infrequent. Catawba.

A. CORDIFOLIUS, L.
Frequent.

A. CORYMBOSUS, Ait.
Florence and Milan; scarce.

A. DIFFUSUS, Ait.
Frequent and variable.

A. DUMOSUS, L.*
Sandy soil, Milan, southern Perkins; infrequent. Oxford; frequent? Flowers white.

A. ERICOIDES, L.
Common on rocky shores.

A. ERICOIDES PLATYPHYLLUS, Torr & Gray.*
Castalia; rare.

A. JUNCEUS, Ait.*

Castalia and east of Milan; scarce.

A. LAEVIS, L.

Milan, Huron, Oxford, Margaretta, Florence, Catawba; rather frequent.

A. MACROPHYLLUS, L.

Frequent but not observed near Sandusky. Put-in-Bay.

A. MULTIFLORUS, Ait.

Dry soil in the limestone region; frequent. Put-in-Bay.

A. NOVÆ-ANGLIÆ, L.

Along roads near Sandusky and south next to the most common Aster. Not so common in the eastern part of the county and on the Peninsula. Kelley's Island, Put-in-Bay; scarce.

A. PANICULATUS, Lam.

Our most common Aster.

A. POLYPHYLLUS, Willd.

Marblehead, Put-in-Bay, Gibraltar, and probably other islands.

A. PRENANTHOIDES, Muhl.

Perkins, Bloomingville, Milan, Berlin, Florence; infrequent.

A. PTARMICOIDES, Torr & Gray.*

Marblehead; local.

A. PUNICEUS, L.

Castalia, Bloomingville, Milan, Florence; infrequent.

A. PUNICEUS LUCIDULUS, Gray.*

Castalia, along the mill race.

A. SAGITTIFOLIUS, Willd.

Common.

A. SALICIFOLIUS, Ait.

Oxford, Milan, Groton, Margaretta, Sandusky, Catawba; infrequent. Many specimens of A. *paniculatus* approach this species.

A. SHORTII, Hook.
Peninsula and Islands; common. Huron and Vermillion Rivers; frequent.

A. TRADESCANTI, L.
Frequent, at least in Perkins and Oxford. Kelley's Island.

A. UMBELLATUS, Mill.
Infrequent.

A. VIMINEUS, Lam.*
Perkins and probably elsewhere.

BIDENS, L. Bur-Marigold.

B. BECKII, Torr.*   Water Marigold.
Black Channel, Biemiller's Cove, East Harbor; scarce.

B. BIPINNATA, L.   Spanish Needles.
Sandusky, Cedar Point, Catawba, North Bass; rare.

B. CERNUA, L.   Smaller Bur-Marigold.
Perkins and Margaretta; scarce.

B. CHRYSANTHEMOIDES, Michx.   Larger Bur-Marigold.
Frequent. Islands.

B. CONNATA, Muhl.   Swamp Beggar-ticks.
Common. One specimen seven feet tall. Some specimens have the awns upwardly barbed.

B. CONNATA COMOSA, Gray.
Frequent.

B. FRONDOSA, L.   Common Beggar-ticks. Stick-tight.
Common. A troublesome weed.

BOLTONIA, L'Her.

B. ASTEROIDES, L'Her.
Sheltered beaches of Lake Erie and Sandusky Bay especially Johnson's Island and the beach stretching from Port Clinton towards Catawba. Not on rocks nor pure sand. Put-in-Bay the only island in the lake.

CALENDULA, L. Marigold.

C. *officinalis*, L.   Garden Marigold.

Sandusky and Put-in-Bay; spreading and escaping, but seldom far from gardens. Hardly naturalized.

CENTAUREA, L.

C. *cyanus*, L.   Blue-bottle.   Corn-flower.

Kelley's Island and elsewhere; sparingly escaped.

CHRYSANTHEMUM, L.

C. *balsamita*, L.   Costmary.

Escaped from gardens in several places.

C. *leucanthemum*, L.   Ox-eye or White Daisy.   White-weed.

Common in several places but not generally distributed.   Put-in-Bay.

C. *parthenium*, Bernh.   Feverfew.

Escaped to waste places in Sandusky and well established in woods on Put-in-Bay.

CNICUS, L.

C. ALTISSIMUS, Willd.

Infrequent.   Kelley's Island.

C. *arvensis*, Hoffm.   Canada Thistle.

Frequent, especially near the Lake and Bay. Islands.

C. DISCOLOR, Muhl.

Frequent.

C. *lanceolatus*, Willd.   Common Thistle.

Common.

C. MUTICUS, Ell.   Swamp Thistle.

Infrequent.

COREOPSIS, L. Tickseed.

C. ARISTOSA, Michx.

Castalia and Venice marshes; common. Cedar Point, Catawba, Vermillion; frequent.

C. DISCOIDEA, Torr & Gray.
Sandusky, Cedar Point, Oxford; locally plentiful.

C. TRICHOSPERMA. Michx. Tickseed Sunflower.
Infrequent.

C. TRICHOSPERMA TENUILOBA, Gray.
Frequent, especially on wet prairies. Kelley's Island. Hundreds of acres of marsh near Bay Bridge glow in autumn with the yellow blossoms, a sight worth going far to see.

C. TRIPTERIS, L. Tall Coreopsis.
Frequent from Milan and Cedar Point west. Peninsula.

ECLIPTA, L.

E. ERECTA, L. (E. ALBA Hassk.)
Sandusky, East Harbor, Lockwood's; scarce.

ERECHTITES, Raf. Fireweed.

E. PRÆALTA, Raf. (E. HIERACIFOLIA, Raf.)
Common.

ERIGERON, L. Fleabane.

E. ANNUUS, Pers. Daisy Fleabane. Sweet Scabious.
Common.

F. BELLIDIFOLIUS, Muhl. Robin's Plantain.
Milan, Perkins, Margaretta Ridge; infrequent.

E. CANADENSIS, L. Horse-weed. Butter-weed.
Common.

E. PHILADELPHICUS, L. Common Fleabane.
Common.

E. STRIGOSUS, Muhl. Daisy Fleabane.
Frequent or common. Islands. Abundant on Marblehead.

EUPATORIUM, L. Thoroughwort.

E. AGERATOIDES, L. White Snakecroot.
Common. Rattlesnake the only island. This plant H. H. Lockwood says is the "Trembleweed" and the cause of milk sickness.

E. ALTISSIMUM, L.
Northwestern Margaretta; infrequent. Johnson's, Marblehead; rare.

E. PERFOLIATUM, L. Thoroughwort. Boneset.
Common.

F. PURPUREUM, L. Joe-Pye Weed. Trumpet-Weed.
Common. Not on the Islands.

E. SESSILIFOLIUM, L. Upland Boneset.
Milan, Huron, Catawba; rare.

GNAPHALIUM, L. Cudweed.

G. DECURRENS, Ives. Everlasting.
Catawba and Florence; very rare.

G. OBTUSIFOLIUM, L. (G. POLYCEPHALUM, Michx.)
Common Everlasting.
Common.

G. PURPUREUM, L. Purplish Cudweed.
Infrequent.

G. ULIGINOSUM, L. Low Cudweed.
Infrequent.

HELENIUM, L. Sneeze-weed.

H. AUTUMNALE, L.
Common at Sandusky and vicinity. Florence.
Catawba.

HELIANTHUS, L. Sunflower.

H. ANNUUS, L.
Frequently escaped. "Cedar Point, far from any house" Ralph H. McKelvey.

H. DECAPETALUS, L.
Frequent.

H. DIVARICATUS, L.
Frequent, especially on Marblehead and the Islands.

H. GIGANTEUS, L.

    Sandusky to Milan and west; common. The so-
called variety, *ambiguus*, occurs in Perkins and
Oxford, and near Port Clinton.

H. GROSSE-SERRATUS, Martens.

    Oxford, Groton, Margaretta; frequent.

H. HIRSUTUS, Raf.

    Cedar Point, Peninsula, Oxford, Margaretta,
Groton; rather common.

H. MOLLIS, Lam.*

    Prairie, Oxford and Huron; enough to supply the
botanists of the world.

H. OCCIDENTALIS, Riddell.

    Castalia cemetery and Kimball; scarce.

H. PARVIFLORUS, Bernh.

    Frequent.

H. STRUMOSUS MOLLIS, Torr & Gray.*

    Oxford, Groton, Castalia, Cedar Point, Port
Clinton; infrequent. Apparently all our specimens of *H. strumosus* are of this variety.

H. TRACHELIIFOLIUS, Willd.

    Florence, Port Clinton; scarce?

H. TUBEROSUS, L.   Jerusalem Artichoke.

    Frequent. Kelley's Island. Put-in-Bay.

HELIOPSIS, Pers. Ox-eye.

H. LÆVIS, Pers.

    Common.

H. SCABRA, Dunal.

    Rather frequent.

INULA, L. Elecampane.

I. *helenium*, L.

    Infrequent. Florence; frequent.

KUHNIA, L.

K. EUPATORIOIDES, L.
Dry soil near Castalia; locally common. San-
✻ dusky and Perkins; scarce.

LEPACHYS, Raf.

L. PINNATIFIDA, Raf.
Common on prairies.

LIATRIS, Schreb. Button Snakeroot.

L. SCARIOSA, Willd.
Catawba, Cedar Point, Margaretta Ridge, southern Perkins, Kimball; plentiful in some places.

L. SPICATA, Willd.
Castalia prairie; abundant and showy. Marblehead, Cedar Point, Oxford, southern Perkins,
Groton, east of Milan; frequent on undisturbed
damp ground.

L. SQUARROSA INTERMEDIA, DC.* Blazing-Star.
Castalia and Sandhill cemeteries.

POLYMNIA, L. Leaf-Cup.

P. CANADENSIS, L.
Cedar Point, Peninsula, Islands; infrequent.

RUDBECKIA, L. Cone-flower.

R. HIRTA, L.
Common. Not on the Islands.

R. LACINIATA, L.
Frequent.

R. TRILOBA, L.
"Port Clinton" Wm. Krebs.

SENECIO, L. Groundsel.

S. ATRIPLICIFOLIUS, Hook. (CACALIA ATRIPLICIFOLIA,
L.) Pale Indian Plantain.
Vermillion River, Florence; frequent. Elsewhere
infrequent.

S. AUREUS, L.　Golden Ragwort.
Frequent.

S. AUREUS OBOVATUS, Torr & Gray.　Squaw-weed.
Common.　Kelley's the only island.

S. AUREUS BALSAMITÆ, Torr & Gray.
Castalia, Perkins, Marblehead, Catawba; frequent.　Put-in-Bay.

SILPHIUM, L.　Rosin-weed.

S. PERFOLIATUM, L.　Cup-Plant.
Huron and Vermillion rivers; infrequent.　Castalia; local.

S. TEREBINTHENACEUM, Jacq.　Prairie Dock.
Common on the prairies.

S. TRIFOLIATUM, L.
Frequent.

SOLIDAGO, L.　Golden-rod.

S. BICOLOR, L.
Frequent.

S. BICOLOR CONCOLOR, Torr & Gray.
Rocky shores of the Put-in-Bay Islands; infrequent.

S. CAESIA, L.
Common in rich woods.　Islands.

S. CANADENSIS, L.
Abundant.

S. JUNCEA, Ait.
Frequent.

S. LANCEOLATA, L.
Common.

S. LATIFOLIA, L.
Florence; frequent.　Vermillion, Berlin Heights, Milan, Perkins, Catawba, Kelley's Island, Green Island, Rattlesnake; scarce.

S. NEMORALIS, Ait.
Frequent. Islands.

S. OHIOENSIS, Riddell.
Castalia prairie; infrequent.

S. PATULA, Muhl.
Florence, Milan, Castalia, Kelley's Island;
infrequent.

S. RIDDELLII, Frank.
Castalia; frequent. Marblehead, Groton, House's
swamp, Perkins; infrequent.

S. RIGIDA, L.
Marblehead and Oxford; frequent. Huron, Sandusky, Margaretta, Groton, Middle Bass; local.

S. RUGOSA, Mill.
East of Milan; local.

S. SEROTINA, Ait.
Frequent.

S. SEROTINA GIGANTEA, Gray.
Milan, Oxford, southern Perkins; scarce.

S. SPECIOSA, Nutt.
Huron River and Peninsula; infrequent. Southern Perkins; scarce.

S. SPECIOSA ANGUSTATA, Torr & Gray.*
Leonard's Hazel Patch, Perkins; rare.

S. TENUIFOLIA, Pursh.
Oxford prairie; abundant.

S. ULMIFOLIA, Muhl.
Marblehead; frequent. Elsewhere infrequent.
Islands.

TANACETUM, L. Tansy.

T. *vulgare*, L.
Roadsides; frequent. Islands. The ordinary
form is the variety *crispum*, but the other occurs
in "Perkins" and on "Kelley's Island."

VERNONIA, Schreb. Iron-weed.

V. ALTISSIMA, Nutt.
   Common.   Kelley's the only island.
V. ALTISSIMA GRANDIFLORA, Nutt.
   Huron, Willow Point, Kelley's Island; infrequent.
V. FASCICULATA, Michx.
   Prairies; frequent.

XANTHIUM, L. Cocklebur.

X. CANADENSE, Mill.
   Common.   The so-called variety *echinatum* is the
   common form near the Bay and Lake.

CORRECTIONS.

Page 7. For Hypericum kalmianum read *Potentilla fruticosa*. The two grow together on the prairie but the latter is more abundant and to it belong the small petrified leaves collected.

Page 28. The four names at the top of first column should be at the bottom.

Page 50. For P. *annual* read P. *annua*.

Page 54. For *hedunculata* read *pedunculata*.

Page 63. For J. TENVIS read J. TENVIS.

Page 76. Place a mark of doubt —?— after occurs, at end of third line.

Page 84. For **AMONACEÆ** read **ANONACEÆ**.

Page 94. For **SAXIFRAGACÆ** read **SAXIFRAGACEÆ**.

Page 150. For T. pratensis read T. *pratensis*.

INDEX.

*** The names of families are in capitals. In the catalogue the genera of each family and the species of each genus are arranged alphabetically.